SpringerBriefs in Electrical and Computer Engineering

Computational Electromagnetics

W0079675

Series editor

Rakesh Mohan Jha, Bangalore, India

More information about this series at http://www.springer.com/series/13885

Hema Singh · R. Chandini
Rakesh Mohan Jha

RCS Estimation of Linear and Planar Dipole Phased Arrays: Approximate Model

 Springer

Hema Singh
Centre for Electromagnetics
CSIR-National Aerospace Laboratories
Bangalore, Karnataka
India

Rakesh Mohan Jha
Centre for Electromagnetics
CSIR-National Aerospace Laboratories
Bangalore, Karnataka
India

R. Chandini
Centre for Electromagnetics
CSIR-National Aerospace Laboratories
Bangalore, Karnataka
India

ISSN 2191-8112   ISSN 2191-8120 (electronic)
SpringerBriefs in Electrical and Computer Engineering
ISSN 2365-6239   ISSN 2365-6247 (electronic)
SpringerBriefs in Computational Electromagnetics
ISBN 978-981-287-753-6   ISBN 978-981-287-754-3   (eBook)
DOI 10.1007/978-981-287-754-3

Library of Congress Control Number: 2015947949

Springer Singapore Heidelberg New York Dordrecht London

Printed on acid-free paper

Springer Science+Business Media Singapore Pte Ltd. is part of Springer Science+Business Media
(www.springer.com)

To Professor R. Narasimha

In Memory of Dr. Rakesh Mohan Jha
Great scientist, mentor, and excellent
human being

Dr. Rakesh Mohan Jha was a brilliant contributor to science, a wonderful human being, and a great mentor and friend to all of us associated with this book. With a heavy heart we mourn his sudden and untimely demise and dedicate this book to his memory.

Foreword

National Aerospace Laboratories (NAL), a constituent of the Council of Scientific and Industrial Research (CSIR), is the only civilian aerospace R&D Institution in India. CSIR-NAL is a high-technology institution focusing on various disciplines in aerospace and has a mandate to develop aerospace technologies with strong science content, design, and build small- and medium-sized civil aircraft prototypes, and support all national aerospace programs. It has many advanced test facilities including trisonic wind tunnels, which are recognized as National Facilities. The areas of expertise and competencies include computational fluid dynamics, experimental aerodynamics, electromagnetics, flight mechanics and control, turbomachinery and combustion, composites for airframes, avionics, aerospace materials, structural design, analysis, and testing. CSIR-NAL is located in Bangalore, India, with the CSIR Headquarters being located in New Delhi.

CSIR-NAL and Springer have recently signed a cooperation agreement for the publication of selected works of authors from CSIR-NAL as Springer book volumes. Within these books, recent research in the different fields of aerospace that demonstrate CSIR-NAL's outstanding research competencies and capabilities to the global scientific community will be documented.

The first set of five books are from selected works carried out at the CSIR-NAL's Centre for Electromagnetics and are presented as part of the series SpringerBriefs in Computational Electromagnetics, which is a sub-series of SpringerBriefs in Electrical and Computer Engineering.

CSIR-NAL's Centre for Electromagnetics mainly addresses issues related to electromagnetic (EM) design and analysis carried out in the context of aerospace engineering in the presence of large airframe structures, which is vastly different and in contrast to classical electromagnetics and which often assumes a free-space ambience. The pioneering work done by the Centre for Electromagnetics in some of these niche areas has led to founding the basis for contemporary theories. For example, the geodesic constant method (GCM) proposed by the scientists of the Centre for Electromagnetics is immensely popular with the peers worldwide, and forms the basis for modern conformal antenna array theory.

The activities of the Centre for Electromagnetics consist of (i) Surface modeling and ray tracing, (ii) Airborne antenna analysis and siting (for aircraft, satellites and SLV), (iii) Radar cross section (RCS) studies of aerospace vehicles, including radar absorbing materials (RAM) and structures (RAS), RCS reduction, and active RCS reduction, (iv) Phased antenna arrays, conformal arrays, and conformal adaptive array design, (v) Frequency-selective surface (FSS), (vi) Airborne and ground-based radomes, (vii) Metamaterials for aerospace applications including in the Terahertz (THz) domain, and (viii) EM characterization of materials.

It is hoped that this dissemination of information through these SpringerBriefs will encourage new research as well as forge new partnerships with academic and research organizations worldwide.

Shyam Chetty
Director
CSIR-National Aerospace Laboratories
Bangalore, India

Preface

The scattering characteristics of phased arrays depend on the reflections a signal undergoes while traveling through the antenna array system including the feed network. One of the approaches being used is to trace the signal path step-by-step and calculate the reflection and the transmission coefficients at each component level and then coherently sum the contributions to arrive at array RCS. For large phased arrays and feed network with multiple coupler levels, the computations become too complicated. In order to avoid such complexity, in this book, radar cross section (RCS) of a parallel-fed linear and planar dipole array is derived using an approximate method. The RCS is expressed in terms of array factor, neglecting phase terms. The impinging signal travels through the antenna system passing through radiators, phase shifters, couplers before reaching the receive port of the feed network. It undergoes reflection and transmission at various levels of feed network due to impedance mismatches. The mutual coupling effect is included in the RCS formulation. The dependence of the RCS pattern on the design parameters, viz., antenna elements, geometric configuration, inter-element spacing, beam scan angle, and components like phase shifters, couplers, terminating impedance is analyzed.

This book presents a detailed formulation for RCS of dipole arrays along with parametric analysis. It provides an insight for graduate and research students, scientists, and academicians for estimation and optimization of RCS of phased arrays with other types of antenna elements in arbitrary geometrical configuration.

Hema Singh
R. Chandini
Rakesh Mohan Jha

Acknowledgments

We would like to thank Mr. Shyam Chetty, Director, CSIR-National Aerospace Laboratories, Bangalore for his permission and support to write this SpringerBrief.

We would also like to acknowledge valuable suggestions from our colleagues at the Centre for Electromagnetics, Dr. R.U. Nair, Dr. Shiv Narayan, Dr. Balamati Choudhury, and Mr. K.S. Venu and their invaluable support during the course of writing this book. We would like to thank Mr. Harish S. Rawat, Ms. Neethu P.S., Mr. Umesh V. Sharma, and Mr. Bala Ankaiah, the project staff at the Centre for Electromagnetics, for their consistent support during the preparation of this manuscript.

But for the concerted support and encouragement from Springer, especially the efforts of Suvira Srivastav, Associate Director, and Swati Mehershi, Senior Editor, Applied Sciences & Engineering, it would not have been possible to bring out this book within such a short span of time. We very much appreciate the continued support by Ms. Kamiya Khatter and Ms. Aparajita Singh of Springer toward bringing out this brief.

Hema Singh
R. Chandini
Rakesh Mohan Jha

Contents

About the Authors

Dr. Hema Singh is currently working as Senior Scientist in Centre for Electromagnetics of CSIR-National Aerospace Laboratories, Bangalore, India. Earlier, she was Lecturer in EEE, BITS, Pilani, India during 2001–2004. She obtained her Ph.D. degree in Electronics Engineering from IIT-BHU, Varanasi India in 2000. Her active area of research is Computational Electromagnetics for Aerospace Applications. More specifically, the topics she has contributed to, are GTD/UTD, EM analysis of propagation in an indoor environment, phased arrays, conformal antennas, radar cross section (RCS) studies including Active RCS Reduction. She received Best Woman Scientist Award in CSIR-NAL, Bangalore for period 2007–2008 for her contribution in the areas of phased antenna array, adaptive arrays, and active RCS reduction. Dr. Singh has co-authored one book, one book chapter, and over 120 scientific research papers and technical reports.

R. Chandini obtained her B.E. (ECE) degree from Visvesvaraya Technological University, Karnataka. She was a Project Engineer at the Centre for Electromagnetics of CSIR-National Aerospace Laboratories, Bangalore, where she worked on RCS studies and conformal arrays.

Dr. Rakesh Mohan Jha was Chief Scientist & Head, Centre for Electromagnetics, CSIR-National Aerospace Laboratories, Bangalore. Dr. Jha obtained a dual degree in BE (Hons.) EEE and M.Sc. (Hons.) Physics from BITS, Pilani (Raj.), India, in 1982. He obtained his Ph.D. (Engg.) degree from Department of Aerospace Engineering of Indian Institute of Science, Bangalore in 1989, in the area of computational electromagnetics for aerospace applications. Dr. Jha was a SERC (UK) Visiting Post-Doctoral Research Fellow at University of Oxford, Department of Engineering Science in 1991. He worked as an Alexander von Humboldt Fellow at the Institute for High-Frequency Techniques and Electronics of the University of Karlsruhe, Germany (1992–1993, 1997). He was awarded the Sir C.V. Raman Award for Aerospace Engineering for the Year 1999. Dr. Jha was elected Fellow of INAE in 2010, for his contributions to the EM Applications to Aerospace

Engineering. He was also the Fellow of IETE and Distinguished Fellow of ICCES. Dr. Jha has authored or co-authored several books, and more than five hundred scientific research papers and technical reports. He passed away during the production of this book of a cardiac arrest.

List of Figures

RCS Estimation of Linear and Planar Dipole Phased Arrays: Approximate Model

Abstract The signal propagation within the phased array system decides the radar cross section (RCS) of phased array. The reflection and transmission coefficients for a signal at different levels of the phase in scattering array system depend on the impedance mismatch and the design parameters. Moreover, the mutual coupling effect in between the antenna elements is an important factor analysis. A phased array system comprises of radiating elements followed by phase shifters, couplers, and terminating load impedance. These components pose respective impedance toward the incoming signal that travels through them before reaching the receive port of the array system. In this book, the RCS of a parallel-fed linear and planar dipole array is derived using an approximate method. The RCS is approximated in terms of array factor, neglecting phase terms. The mutual coupling effect is taken into account. The dependence of the RCS pattern on the design parameters is analyzed. The approximate model employed proves to be an efficient method for RCS estimation of phased arrays. This book presents a detailed formulation of approximate method to determine radar cross section of phased arrays. It explains the RCS estimation of phased array using schematics and illustrations. This book helps readers to understand the impinging signal path and its reflections/transmissions within the phased array system.

Keywords Radar cross section · Dipole phased array · Approximate model · Mutual coupling · Impedance mismatch · Scattered field · Radiator · Phase shifter · Coupler · Terminating load

1 Introduction

The reflections and transmission of the impinging signal within the phased array system decides the radar cross section (RCS) of phased array. The corresponding reflection and transmission coefficients for a signal at different levels of the phased array system depend on the impedance mismatch and the design parameters. Moreover the mutual coupling effect in between the antenna elements is an important

© The Author(s) 2016   1
H. Singh et al., *RCS Estimation of Linear and Planar Dipole Phased Arrays: Approximate Model*, SpringerBriefs in Computational Electromagnetics, DOI 10.1007/978-981-287-754-3_1

factor in scattering analysis. A phased array system comprises of radiating elements followed by phase shifters, couplers, and terminating load impedance. These components pose respective impedance towards the incoming signal that travels through the components before reaching receive port of the array system.

This book aims at the estimation and reduction of in-band RCS of phased antenna arrays considering only the antenna mode scattering, which is dominant as compared to that of structural mode for an in-band operating stand-alone antenna array. In order to achieve this objective, an efficient and accurate RCS model, which can optimize the trade-off between antenna RCS and radiation performance is a crucial requirement.

In this book, RCS of parallel-fed dipole array is determined using approximate model. The analytical formulation is presented for both linear and planar dipole array. The scattering of the impinging signal till second level of couplers is considered. The mutual coupling effect is included. The self and mutual impedance of the array are calculated using the finite dimensions of dipole element. The high order reflections and transmissions and edge effect are ignored. The overall RCS of dipole array is calculated by approximating the individual scattering contributions at different levels in terms of array factors, neglecting the phase terms. The array RCS depends on the array design parameters, viz. antenna elements, geometric configuration, inter-element spacing, components, viz. phase shifters, couplers, terminating impedance, and the feed configuration (Sneha et al. 2013a, b).

2 RCS of Parallel-Fed Linear Dipole Array Using Approximate Model

The array RCS is the ratio of the scattered power to the incident power, given by (Jenn 1995)

$$\sigma(\theta_i) = \lim_{R \to \infty} 4\pi R^2 \frac{\left|\vec{E}_s\right|^2}{\left|\vec{E}^i\right|^2} \tag{1}$$

where, the scattered field $\vec{E}^s = \sum_m \vec{E}^s_m$.

The scattering from mth antenna element of array is expressed as (Sneha et al. 2012)

$$\vec{E}^s_m(\theta, \phi) = \left[\frac{j\eta}{4_{a_m}} \vec{h}\left\{\vec{h} \cdot \vec{E}^i(\theta, \phi)\right\} \frac{e^{-jk\vec{R}}}{R}\right] \Gamma^r_m(\theta, \phi) \tag{2}$$

where, $\vec{E}^i$ is the incident field, (θ, ϕ) is direction of the incident wave, λ is the wavelength of the impinging signal, η is the free space impedance, k is the wave

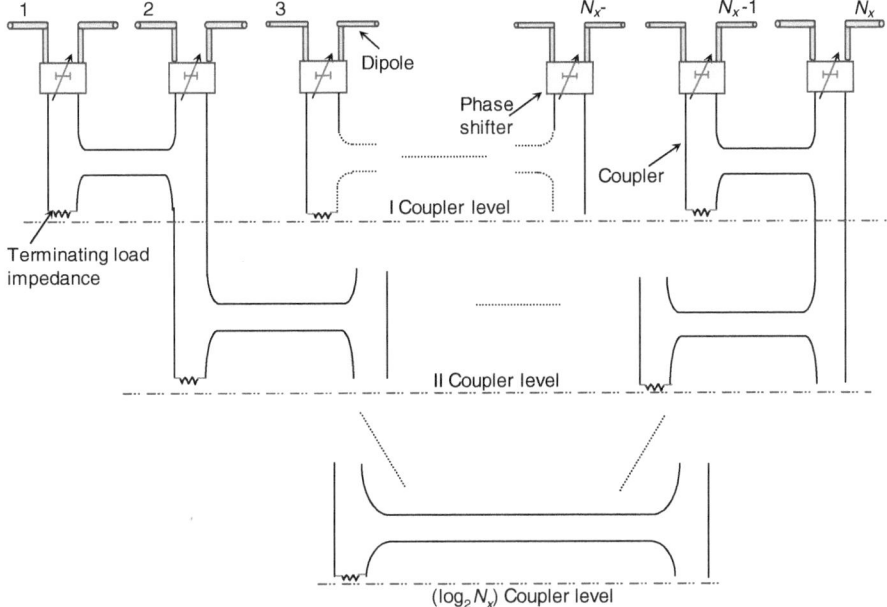

Fig. 1 A parallel-fed linear dipole array

number, R is the far-field, Γ_m^r is the reflection coefficient of mth antenna element, $\vec{h}$ is the effective height of the element.

$Z_{a_m} = R_{a_m} + jX_{a_m}$ is the radiation impedance of mth antenna element. Here R_{a_m} and X_{a_m} are the dipole antenna resistance and reactance respectively.

Figure 1 shows the schematic of a typical linear N_x-lement dipole array with parallel feed network. It can be seen that the feed network consists of phase shifters, different levels of couplers and terminating load impedance.

The total scattered field of linear dipole array is obtained by summing over N_x antenna elements.

$$\vec{E}^s(\theta, \phi) = \sum_{m=1}^{N_x} \vec{E}_m^s(\theta, \phi) = \sum_{m=1}^{N_x} \left[\left\{ \frac{j\eta_o}{4\lambda(R_{r_m} + jX_{a_m})} \atop \times h^2 \cdot \cos\theta \cdot \vec{E}_m^r(\theta, \phi) \right\} \frac{e^{-jk\vec{R}}}{R} \hat{x} \right] \quad (3)$$

where h is the effective height of antenna element, given by $\vec{h} = h\hat{x} = \left(\int_{\Delta l} \cos(kl)dl \right)\hat{x}$, l being the dipole length.

It is to be noted that $R_{a_m} = R_{r_m} + R_{d_m}$, where R_{r_m} is the radiation resistance and R_{d_m} is the loss resistance of the mth antenna element, taken as zero.

This yields the total RCS of a linear dipole array with N_x elements, given by (Sneha et al. 2014)

$$\sigma(\theta, \phi) = \lim_{R \to \infty} 4\pi R^2 \sum_{m=1}^{N_x} \left[\left\{ \frac{j\eta_o}{4\lambda(R_{r_m} + jX_{a_m})} h^2 \cdot \cos\theta \cdot \vec{E}_m^r(\theta, \phi) \right\} \frac{e^{-j\vec{k}\vec{R}}}{R} \hat{x} \right]^2$$

$$= 4\pi \sum_{m=1}^{N_x} \left[\frac{j\eta_o}{4\lambda(R_{r_m} + jX_{a_m})} h^2 \cdot \cos\theta \cdot \vec{E}_m^r(\theta, \phi) \right]^2 \qquad (4)$$

$$\vec{E}_m^r(\theta, \phi) = \Gamma_m^r(\theta, \phi) \cdot E^i = \Gamma_m^r(\theta, \phi) \cdot 1 \cdot e^{j(m-1)\alpha} \hat{\theta}, \qquad (5)$$

where $\alpha = kd_x \sin\theta \cos\phi$ is the inter-element delay.

As mentioned above, the feed network consists of radiators, phase shifters, couplers and terminating load. Thus for estimating the total array RCS, the scattering from each component is to be considered.

Radiators The RCS of the radiators is given by (Sneha et al. 2012)

$$\sigma_r(\theta, \phi) = \frac{4\pi}{\lambda^2} \left| F \sum_{m=1}^{N_x} \vec{E}_{r_m}^r(\theta, \phi) = F \sum_{m=1}^{N_x} \Gamma_{r_m} e^{j2(m-1)\alpha} \right|^2 \qquad (6)$$

Using approximate model (Jenn and Flokas 1996) the scattered field is expressed in terms of array factors. The phase factors are ignored.

Thus, the RCS of radiators can be expressed as

$$\sigma_r(\theta, \phi) = \frac{4\pi}{\lambda^2} \left| F \sum_{m=1}^{N_x} \vec{E}_{r_m}^r(\theta, \phi) \right|^2 = \frac{4\pi}{\lambda^2} \left| F \sum_{m=1}^{N_x} \Gamma_{r_m} e^{j2(m-1)\alpha} \right|^2$$

$$\approx \frac{4\pi}{\lambda^2} \left| F \left[\sum_{m=1}^{N_x} e^{j2(m-1)\alpha} \right] \left[\sum_{m=1}^{N_x} \Gamma_{r_m} \right] \right|^2 \qquad (7)$$

$$\text{or } \sigma_r(\theta, \phi) = \frac{4\pi}{\lambda^2} \left| F \left[\sum_{m=1}^{N_x} \Gamma_{r_m} \right] \left[\sum_{p=0}^{N_x} e^{j2p\alpha} \right] \right|^2 ; \; p = m - 1 \qquad (7a)$$

Using closed-form array theory (Jenn 1995) gives

$$\sigma_r(\theta, \phi) = \frac{4\pi}{\lambda^2} \left| F \left[\sum_{m=1}^{N_x} \Gamma_{r_m} \right] \left[\sum_{p=0}^{N_x} e^{j2p\alpha} \right] \right|^2 = \frac{4\pi}{\lambda^2} \left| F \left(\frac{1 - e^{2j\alpha N_x}}{1 - e^{2j\alpha}} \right) \sum_{m=1}^{N_x} \Gamma_{r_m} \right|^2 \qquad (7b)$$

$$\sigma_r(\theta, \phi) = \frac{4\pi}{\lambda^2} \left| F \frac{\sin(N_x \alpha_x)}{\sin \alpha_x} e^{j\alpha(N_x - 1)} \sum_{n=1}^{N_x} \Gamma_{r_m} \right|^2 \qquad (7c)$$

Normalizing w.r.t. N_x, the RCS of radiators is given by

$$\sigma_r(\theta, \phi)|_{\text{normalized}} = \frac{4\pi}{\lambda^2} \left| F \frac{\sin(N_x \alpha_x)}{N_x \sin \alpha_x} e^{j\alpha(N_x - 1)} \sum_{m=1}^{N_x} \Gamma_{r_m} \right|^2 \qquad (7d)$$

$$\text{where } F = \frac{j\eta_o}{4\lambda Z_{a_m}} \left(\int_{\Delta l} \cos(kl) dl \right)^2 \cos\theta \qquad (7e)$$

Neglecting the phase terms, one gets

$$\sigma_r(\theta, \phi)|_{\text{normalized}} = \frac{4\pi}{\lambda^2} \left| F \frac{\sin(N_x \alpha_x)}{N_x \sin \alpha_x} \sum_{m=1}^{N_x} \Gamma_{r_m} \right|^2 \qquad (8)$$

$$\text{where } \Gamma_{r_m} = \left| \frac{Z_{a_m} - Z_o}{Z_{a_m} + Z_o} \right| \qquad (8a)$$

The impedance Z_{a_m} is given by

$$Z_{a_{x,y}} = \begin{pmatrix} z_{a_{1,1}} & z_{a_{1,2}} & \cdots & z_{a_{1,N_x}} \\ z_{a_{2,1}} & z_{a_{2,2}} & \cdots & z_{a_{2,N_x}} \\ \vdots & \vdots & \ddots & \vdots \\ z_{a_{N_x,1}} & z_{a_{N_x,2}} & \cdots & z_{a_{N_x,N_x}} \end{pmatrix} \qquad (9)$$

In other words, the total antenna impedance at the mth element is (Elliot 2005)

$$Z_{a_m} = \sum_{m=1}^{N_x} z_{a_{x,y}} \frac{I_y}{I_x} \qquad (10)$$

where, I_n is the current at the feed terminals of mth antenna element. The expressions for self and mutual impedance depend on the array geometric configuration. In this document, side-by-side and parallel-in-echelon dipole arrays are considered.

Phase shifters The phase shifters are modeled as lossless delay line. The RCS contribution from the phase shifters of the parallel feed network is expressed as (Sneha et al. 2012)

$$\sigma_p(\theta, \phi) = \frac{4\pi}{\lambda^2} \left| F \sum_{m=1}^{N_x} \vec{E}_{p_m}^r(\theta, \phi) \right|^2 = \frac{4\pi}{\lambda^2} \left| F \sum_{m=1}^{N_x} T_{r_m}^2 \Gamma_{p_m} e^{j2(m-1)\alpha} \right|^2$$
$$= \frac{4\pi}{\lambda^2} \left| F \left[\sum_{m=1}^{N_x} T_{r_m}^2 \Gamma_{p_m} \right] \left[\sum_{m=1}^{N_x} e^{j2(m-1)\alpha} \right] \right|^2 \qquad (11)$$

$$\text{or } \sigma_p(\theta, \phi) = \frac{4\pi}{\lambda^2} \left| F \left[\sum_{m=1}^{N_x} T_{r_m}^2 \Gamma_{p_m} \right] \left[\sum_{p=1}^{N_x-1} e^{j2p\alpha} \right] \right|^2 \quad ; p = m - 1 \tag{11a}$$

Using approximate model, one gets

$$\sigma_p(\theta, \phi) = \frac{4\pi}{\lambda^2} \left| F \frac{\sin(N_x \alpha_x)}{N_x \sin \alpha_x} \sum_{m=1}^{N_x} T_{r_m}^2 \Gamma_{p_m} e^{j\alpha(N_x-1)} \right|^2 \tag{12}$$

Neglecting the phase terms, it becomes

$$\sigma_p(\theta, \phi)\big|_{\text{normalized}} = \frac{4\pi}{\lambda^2} \left| F \frac{\sin(N_x \alpha_x)}{N_x \sin \alpha_x} \sum_{m=1}^{N_x} T_{r_m}^2 \Gamma_{p_m} \right|^2 \tag{13}$$

where

$$\Gamma_{p_m} = \left| \frac{Z_{p_m} - Z_o}{Z_{p_m} + Z_o} \right| \text{ and } |T_{r_m}|^2 = 1 - |\Gamma_{r_m}|^2 \tag{14}$$

Couplers After phase-shifters in parallel feed network, the signal reaches the input arms of the first level couplers. In parallel feed network, a single coupler is connected to multiple antennas depending upon the coupler level. For example, in first level coupler, two dipole antennas are connected to one coupler where in the second level coupler; four dipole antennas are connected to a single coupler. In general, there will be $N/2^q$ couplers, N is the number of elements and q is the coupler level.

Here the couplers are taken as a four-port lossless device. It has two input port arms (Port 2 and 3), one sum arm (Port 1) and one difference arm (Port 4). The transmission and the coupling coefficients of the couplers are taken as $(T_{c_{qi}}, c_{qi})$, q represents the coupler level in the feed network and i is the coupler number in a given coupler level. In general, the terminating load impedance is connected to the difference port of the coupler. The sum port of the coupler connects to subsequent coupler levels. The magnitude of reflected field, at any junction, depends on the impedance mismatch experienced by the signal during its path from the aperture to the receive port.

The RCS contribution from the couplers of a parallel feed is given by (Sneha et al. 2013c)

$$\sigma_{cp}(\theta, \phi)\big|_{\text{normalized}} = \frac{4\pi}{\lambda^2} \left| F \sum_{m=1}^{N_x} \vec{E}_{cp_m}^r (\theta, \phi) \right|^2$$

$$= \frac{4\pi}{\lambda^2} \left| F \frac{\sin(N_x \xi_x)}{N_x \sin \xi_x} \sum_{m=1}^{N_x} T_{r_m}^2 \Gamma_{cp_m} T_{p_m}^2 \right|^2 \tag{15}$$

where the reflection coefficient of coupler in first level is given by

$$\Gamma_{cp_m} = \left| \frac{Z_{33_{1i}} - Z_{p_m}}{Z_{33_{1i}} + Z_{p_m}} \right| \text{ for odd-numbered elements, i.e., at Port 3} \tag{15a}$$

$$\Gamma_{cp_m} = \left| \frac{Z_{22_{1i}} - Z_{p_m}}{Z_{22_{1i}} + Z_{p_m}} \right| \text{ for even-numbered elements, i.e., at Port 2} \tag{16}$$

$Z_{33_{1i}}$ is the impedance at Port 3, $1i$ indicates the ith coupler in the first level of the feed network; $i = 1, 2, ..., N_x/2$.

The variable i increments for every two elements in the first level couplers. Furthermore the Port 2 and Port 3 of each first level coupler are connected to the adjacent antenna elements. Thus the impedances exhibited by the end terminals of phase-shifters, Z_{p_m} differ at coupler ports 2 and 3.

Scattering at the sum and difference arms of first level couplers Once the signal enters the input arms of the coupler, after reflections it passes through the first level of the couplers. As said above, the coupler has sum and difference arms. The impedance and hence the reflection coefficients associated with the sum and difference arms will be different. Thus, one has

$$\Gamma_{s_{qi}} = \left| \frac{Z_{31_{qi}} - Z_{11_{qi}}}{Z_{31_{qi}} + Z_{11_{qi}}} \right| \quad \text{for odd-arms of the coupler} \tag{17}$$

$$\Gamma_{s_{qi}} = \left| \frac{Z_{21_{qi}} - Z_{11_{qi}}}{Z_{21_{qi}} + Z_{11_{qi}}} \right| \quad \text{for even-arms of the coupler} \tag{18}$$

and

$$\Gamma_{d_{qi}} = \left| \frac{Z_{34_{qi}} - Z_{44_{qi}}}{Z_{34_{qi}} + Z_{44_{qi}}} \right| \quad \text{for odd-arms of the coupler} \tag{19}$$

$$\Gamma_{d_{qi}} = \left| \frac{Z_{24_{qi}} - Z_{44_{qi}}}{Z_{24_{qi}} + Z_{44_{qi}}} \right| \quad \text{for even-arms of the coupler} \tag{20}$$

Since two antenna elements are connected to a single coupler, the RCS contribution due to scattering at sum and difference arms of first level coupler depends on the signal paths at both mth and $(m + 1)$th element. Tracing the signal path, when the signal is incident at mth element is shown in Fig. 2. Similarly the signal path can be traced for $(m + 1)$th element.

Fig. 2 Signal reflections due to the sum and difference arms of first level coupler. **a** Sum port. **b** Difference port

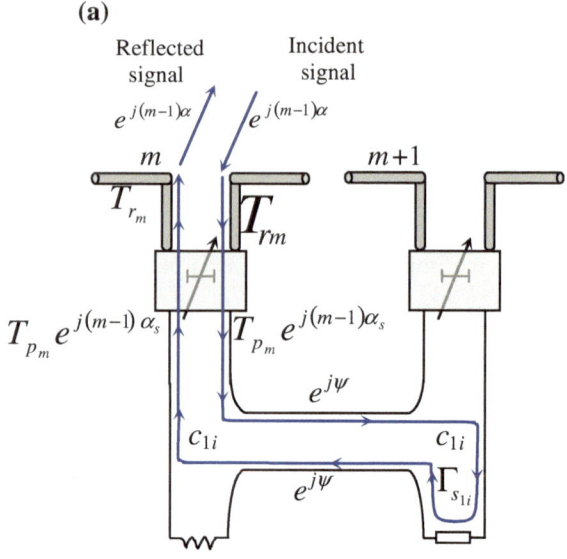

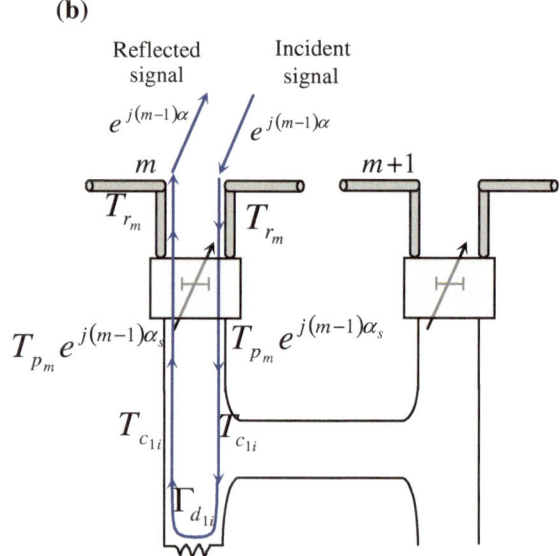

The RCS at mth element due to the sum and difference arms of first level couplers is given by

$$\sigma_{\mathrm{sd}_1}(\theta, \phi)_m|_{\text{normalized}} = \frac{4\pi}{\lambda^2}\left| F \sum_{m=1}^{N_x} \vec{E}_{m_1}^r(\theta, \phi) \right|^2 \tag{21}$$

$$
\sigma_{sd_1}(\theta, \phi)_m |_{\text{normalized}} = \frac{4\pi}{\lambda^2} \left| F \frac{\sin(N_x \xi_x)}{N_x \sin(2\xi_x)} \sum_{m=1,3\ldots}^{N_x-1} T_{r_m} T_{p_m} \left\{ \Gamma_{s_{1i}} c_{1i} e^{j\psi} \begin{pmatrix} c_{1i} e^{j\psi} T_{r_m} T_{p_m} \\ + T_{r_{m+1}} T_{p_{m+1}} T_{c_{1i}} \end{pmatrix} + \Gamma_{d_{1i}} T_{c_{1i}} \begin{pmatrix} T_{r_m} T_{p_m} T_{c_{1i}} \\ + T_{r_{m+1}} T_{p_{m+1}} c_{1i} e^{j\psi} \end{pmatrix} \right\} \right|^2
$$

(21a)

Similarly, the RCS at $(m + 1)$th element due to the sum and difference arms of first level couplers is given by

$$
\sigma_{sd_1}(\theta, \phi)_{(m+1)} |_{\text{normalized}} = \frac{4\pi}{\lambda^2} \left| F \sum_{m=1}^{N_x} \vec{E}^r_{(m+1)_i}(\theta, \phi) \right|^2
$$

$$
= \frac{4\pi}{\lambda^2} \left| F \frac{\sin(N_x \xi_x)}{N_x \sin(2\xi_x)} \sum_{m=1,3\ldots}^{N_x-1} T_{r_{m+1}} T_{p_{m+1}} \left\{ \begin{matrix} \Gamma_{s_{1i}} T_{c_{1i}} \\ \times \left(T_{r_m} T_{p_m} c_{1i} e^{j\psi} + T_{c_{1i}} T_{r_{m+1}} T_{p_{m+1}} \right) \\ + \Gamma_{d_{1i}} c_{1i} e^{j\psi} \\ \times \left(T_{r_m} T_{p_m} T_{c_{1i}} + T_{r_{m+1}} T_{p_{m+1}} c_{1i} e^{j\psi} \right) \end{matrix} \right\} \right|^2
$$

(22)

Thus, the total RCS contribution due to the first level couplers of a parallel feed is given by

$$
\sigma_{sd_1}(\theta, \phi) |_{\text{normalized}} = \sigma_{sd_1}(\theta, \phi)_m |_{\text{normalized}} + \sigma_{sd_1}(\theta, \phi)_{(m+1)} |_{\text{normalized}} \qquad (23)
$$

Scattering till First Level Couplers The total RCS of the dipole array due to the impedance mismatches till first level couplers of a parallel feed network is given by

$$
\sigma(\theta, \phi) |_{\text{normalized}} = \left[\sigma_r(\theta, \phi) |_{\text{normalized}} \right] + \left[\sigma_p(\theta, \phi) |_{\text{normalized}} \right]
$$
$$
+ \left[\sigma_{cp}(\theta, \phi) |_{\text{normalized}} \right] + \left[\sigma_{sd_1}(\theta, \phi) |_{\text{normalized}} \right]
$$

(24)

Scattering at the sum *and difference arms of second level couplers* In second level coupler of the parallel feed network, each coupler is connected to four adjacent dipole antennas in the array. This means that in order to obtain the scattered field at the second level coupler, every fifth element is considered, i.e. $m = 1$, 5, 9, ... N_x. In other words, the scattered field at mth element depends on the reflections from m, $(m + 1)$, $(m + 2)$ and $(m + 3)$ elements. Hence the total scattered field of the dipole array due to the mismatches at second level couplers can be analyzed by dividing the array into sub-arrays, each of $4(=2^2)$ elements.

By tracing the signal path at mth dipole element due to scattering at second level coupler (Sneha et al. 2013c), the RCS contribution according to the approximate model is obtained as follows:

$$
\sigma_{sd_2}(\theta,\phi)_m \big|_{\text{normalized}} = \frac{4\pi}{\lambda^2} \left| F \frac{\sin(N_x \xi_x)}{N_x \sin(4\xi_x)} \sum_{m=1,5\ldots}^{N_x-3} T_{r_m} T_{p_m} c_{1i} e^{j\psi} T_{s_{1i}} \right.
$$

$$
\left. \times \left[\Gamma_{s_{2i'}} c_{2i'} e^{j\psi} \left\{ \begin{array}{l} T_{r_m} T_{p_m} c_{1i} e^{j\psi} T_{s_{1i}} c_{2i'} e^{j\psi} \\ + T_{r_{m+1}} T_{p_{m+1}} T_{c_{1i}} T_{s_{1i}} c_{2i'} e^{j\psi} \\ + T_{r_{m+2}} T_{p_{m+2}} c_{1(i+1)} e^{j\psi} T_{s_{1(i+1)}} T_{c_{2i'}} \\ + T_{r_{m+3}} T_{p_{m+3}} T_{c_{1(i+1)}} T_{s_{1(i+1)}} T_{c_{2i'}} \end{array} \right\} \right. \right.
$$

$$
\left. \left. + \Gamma_{d_{2i'}} T_{c_{2i'}} \left\{ \begin{array}{l} T_{r_m} T_{p_m} c_{1i} e^{j\psi} T_{s_{1i}} T_{c_{2i'}} \\ + T_{r_{m+1}} T_{p_{m+1}} T_{c_{1i}} T_{s_{1i}} T_{c_{2i'}} \\ + T_{r_{m+2}} T_{p_{m+2}} c_{1(i+1)} e^{j\psi} T_{s_{1(i+1)}} c_{2i'} e^{j\psi} \\ + T_{r_{m+3}} T_{p_{m+3}} T_{c_{1(i+1)}} T_{s_{1(i+1)}} c_{2i'} e^{j\psi} \end{array} \right\} \right] \right|^2 \tag{25}
$$

Similarly, the RCS contribution at $(m+1)$th, $(m+2)$th and $(m+3)$th antenna element due to scattering in second level coupler level are given as

$$
\sigma_{sd_2}(\theta,\phi)_{(m+1)} \big|_{\text{normalized}} = \frac{4\pi}{\lambda^2} \left| F \frac{\sin(N_x \xi_x)}{N_x \sin(4\xi_x)} \sum_{m=1,5\ldots}^{N_x-3} T_{r_{m+1}} T_{p_{m+1}} T_{c_{1i}} T_{s_{1i}} \right.
$$

$$
\left. \times \left[\Gamma_{s_{2i'}} c_{2i'} e^{j\psi} \left\{ \begin{array}{l} T_{r_m} T_{p_m} c_{1i} e^{j\psi} T_{s_{1i}} c_{2i'} e^{j\psi} \\ + T_{r_{m+1}} T_{p_{m+1}} T_{c_{1i}} T_{s_{1i}} c_{2i'} e^{j\psi} \\ + T_{r_{m+2}} T_{p_{m+2}} c_{1(i+1)} e^{j\psi} T_{s_{1(i+1)}} T_{c_{2i'}} \\ + T_{r_{m+3}} T_{p_{m+3}} T_{c_{1(i+1)}} T_{s_{1(i+1)}} T_{c_{2i'}} \end{array} \right\} \right. \right.
$$

$$
\left. \left. + \Gamma_{d_{2i'}} T_{c_{2i'}} \left\{ \begin{array}{l} T_{r_m} T_{p_m} c_{1i} e^{j\psi} T_{s_{1i}} T_{c_{2i'}} \\ + T_{r_{m+1}} T_{p_{m+1}} T_{c_{1i}} T_{s_{1i}} T_{c_{2i'}} \\ + T_{r_{m+2}} T_{p_{m+2}} c_{1(i+1)} e^{j\psi} T_{s_{1(i+1)}} c_{2i'} e^{j\psi} \\ + T_{r_{m+3}} T_{p_{m+3}} T_{c_{1(i+1)}} T_{s_{1(i+1)}} c_{2i'} e^{j\psi} \end{array} \right\} \right] \right|^2 \tag{26}
$$

$$\sigma_{\text{sd}_2}(\theta,\phi)_{(m+2)}\big|_{\text{normalized}} = \frac{4\pi}{\lambda^2}\left|F\frac{\sin(N_x\xi_x)}{N_x\sin(4\xi_x)}\sum_{m=1,5\dots}^{N_x-3}T_{r_{m+2}}T_{p_{m+2}}c_{1(i+1)}e^{j\psi}T_{s_{1(i+1)}}\right.$$

$$\times\left[\Gamma_{s_{2i'}}c_{2i'}e^{j\psi}\left\{\begin{array}{l}T_{r_m}T_{p_m}c_{1i}e^{j\psi}T_{s_{1i}}c_{2i'}e^{j\psi}\\+T_{r_{m+1}}T_{p_{m+1}}T_{c_{1i}}T_{s_{1i}}c_{2i'}e^{j\psi}\\+T_{r_{m+2}}T_{p_{m+2}}c_{1(i+1)}e^{j\psi}T_{s_{1(i+1)}}T_{c_{2i'}}\\+T_{r_{m+3}}T_{p_{m+3}}T_{c_{1(i+1)}}T_{s_{1(i+1)}}T_{c_{2i'}}\end{array}\right\}\right.$$

$$\left.\left.+\Gamma_{d_{2i'}}T_{c_{2i'}}\left\{\begin{array}{l}T_{r_m}T_{p_m}c_{1i}e^{j\psi}T_{s_{1i}}T_{c_{2i'}}\\+T_{r_{m+1}}T_{p_{m+1}}T_{c_{1i}}T_{s_{1i}}T_{c_{2i'}}\\+T_{r_{m+2}}T_{p_{m+2}}c_{1(i+1)}e^{j\psi}T_{s_{1(i+1)}}c_{2i'}e^{j\psi}\\+T_{r_{m+3}}T_{p_{m+3}}T_{c_{1(i+1)}}T_{s_{1(i+1)}}c_{2i'}e^{j\psi}\end{array}\right\}\right]\right|^2$$

$$(27)$$

$$\sigma_{\text{sd}_2}(\theta,\phi)_{(m+3)}\big|_{\text{normalized}} = \frac{4\pi}{\lambda^2}\left|F\frac{\sin(N_x\xi_x)}{N_x\sin(4\xi_x)}\sum_{m=1,5\dots}^{N_x-3}T_{r_{m+3}}T_{p_{m+3}}T_{s_{1(i+1)}}T_{c_{1(i+1)}}\right.$$

$$\times\left[\Gamma_{s_{2i'}}c_{2i'}e^{j\psi}\left\{\begin{array}{l}T_{r_m}T_{p_m}c_{1i}e^{j\psi}T_{s_{1i}}c_{2i'}e^{j\psi}\\+T_{r_{m+1}}T_{p_{m+1}}T_{c_{1i}}T_{s_{1i}}c_{2i'}e^{j\psi}\\+T_{r_{m+2}}T_{p_{m+2}}c_{1(i+1)}e^{j\psi}T_{s_{1(i+1)}}T_{c_{2i'}}\\+T_{r_{m+3}}T_{p_{m+3}}T_{c_{1(i+1)}}T_{s_{1(i+1)}}T_{c_{2i'}}\end{array}\right\}\right.$$

$$\left.\left.+\Gamma_{d_{2i'}}T_{c_{2i'}}\left\{\begin{array}{l}T_{r_m}T_{p_m}c_{1i}e^{j\psi}T_{s_{1i}}T_{c_{2i'}}\\+T_{r_{m+1}}T_{p_{m+1}}T_{c_{1i}}T_{s_{1i}}T_{c_{2i'}}\\+T_{r_{m+2}}T_{p_{m+2}}c_{1(i+1)}e^{j\psi}T_{s_{1(i+1)}}c_{2i'}e^{j\psi}\\+T_{r_{m+3}}T_{p_{m+3}}T_{c_{1(i+1)}}T_{s_{1(i+1)}}c_{2i'}e^{j\psi}\end{array}\right\}\right]\right|^2$$

$$(28)$$

Thus, the array RCS due to scattering at second level of couplers is given by

$$\sigma_{\text{sd}_2}(\theta,\phi)\big|_{\text{normalized}} = \left\{\begin{array}{l}\sigma_{\text{sd}_2}(\theta,\phi)_m\big|_{\text{normalized}} + \sigma_{\text{sd}_2}(\theta,\phi)_{(m+1)}\big|_{\text{normalized}}\\+\sigma_{\text{sd}_2}(\theta,\phi)_{(m+2)}\big|_{\text{normalized}} + \sigma_{\text{sd}_2}(\theta,\phi)_{(m+3)}\big|_{\text{normalized}}\end{array}\right\}$$

$$(29)$$

Scattering till second level couplers The total RCS of linear dipole array with parallel feed network due to the mismatches in the feed network till second level couplers is expressed as

$$\sigma(\theta, \phi) = \left[\sigma_r(\theta, \phi)|_{\text{normalized}}\right] + \left[\sigma_p(\theta, \phi)|_{\text{normalized}}\right] + \left[\sigma_{cp}(\theta, \phi)|_{\text{normalized}}\right]$$
$$+ \left[\sigma_{\text{sd}_1}(\theta, \phi)|_{\text{normalized}}\right] + \left[\sigma_{\text{sd}_2}(\theta, \phi)|_{\text{normalized}}\right] \tag{30}$$

3 Linear Dipole Array: Simulation Results

A systematic step-by-step approach has been followed to compute array RCS based on design parameters and mutual coupling effect. The phased arrays consisting of half-wavelength dipole antennas with low gain in their out-band region are considered. This assumption serves to neglect scattering in wide angles and thus nullifies the effect of structural mode RCS of phased array. Further low-gain in out-band region eliminates the antenna mode scattering when there is perfect matching between the dipole elements. This is because antenna mode scattering no more remain dominant in perfect match case, even with large reflection coefficient. However, mismatches within the feed network during hardware fabrication are almost unavoidable and thus antenna mode remains to be a dominant RCS contributor in case of a phased antenna array operating within its frequency band.

All the array elements are assumed to be aligned along x-axis (for linear array) with equal inter-element spacing, *d*. The z-axis is taken as the broadside of the array. The radar frequency and the operating frequency of the antenna array are assumed to be the same.

Figure 3 shows the RCS of 64-element parallel-in-echelon dipole array with parallel feed network ($Z_o = 75\ \Omega$; $Z_l = 50\ \Omega$).

The length and radius of the dipole are taken 0.5 and 0.001λ respectively. The inter-element spacing is 0.4λ; the offset height from the reference plane is taken as 0.25λ. The mutual coupling is included to estimate the scattered field till second level couplers. The specular lobe and the lobes due to mismatches at couplers are clearly visible in the RCS pattern.

Figure 4 demonstrate the role of array configuration on RCS pattern of 64-element linear dipole array with parallel feed network. The load impedance and characteristic impedance are taken as 75 and 50 Ω respectively. The other parameters like dipole length, radius, inter-element spacing are kept same as in Fig. 3. Three configurations, viz. side-by-side, collinear, parallel-in-echelon are considered. It is apparent that the RCS of the parallel-in-echelon configuration is lowest while the collinear configuration has highest RCS value.

Figure 5 presents the role of coupler level in RCS pattern of 64-element linear side-by-side parallel-fed dipole array. All the parameters are kept same. It can be observed that RCS pattern shows extra lobes due to mismatches at second level

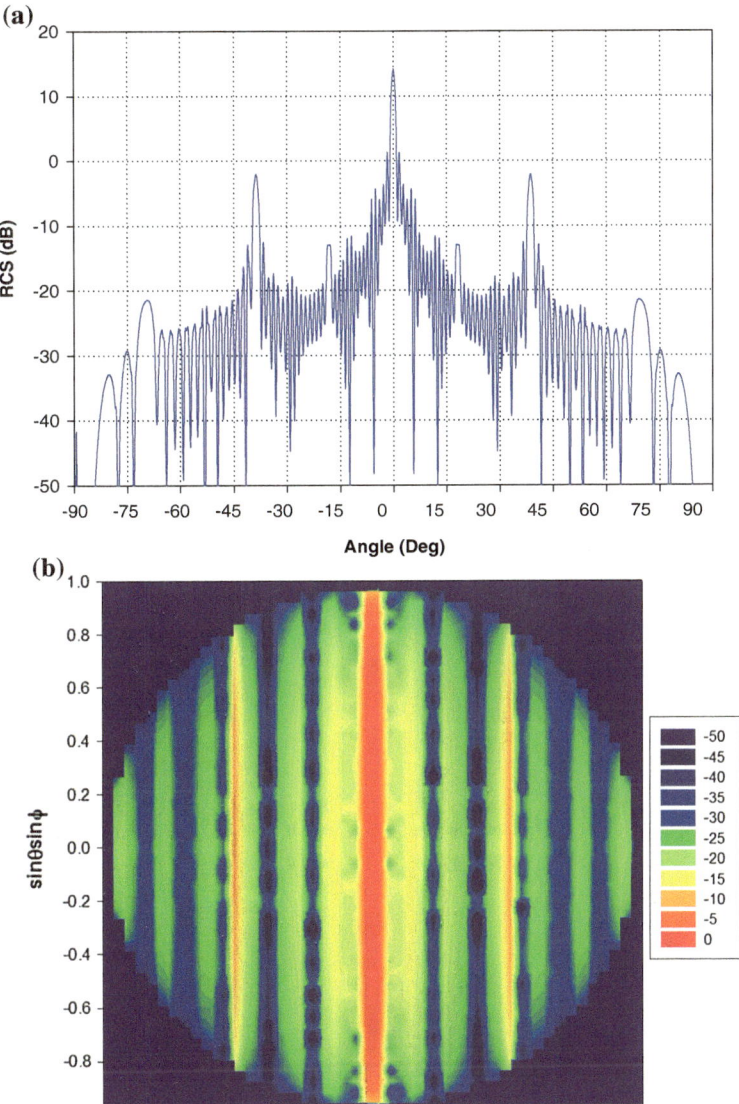

Fig. 3 Broadside RCS of 64-element linear parallel-in-echelon dipole array. Scattering till second level coupler considered. **a** 2-D plot. **b** Contour plot

couplers. As expected the amplitude of the lobes due to second level couplers is less than that of first level coupler lobes. There is no effect on specular lobe RCS.

Next, the effect of beam scanning on the RCS pattern is analyzed (Fig. 6). A 32-element side-by-side linear dipole array is considered. The RCS pattern is calculated for scan angle of 0°, 30°, and 60°. It is apparent that as beam scans the

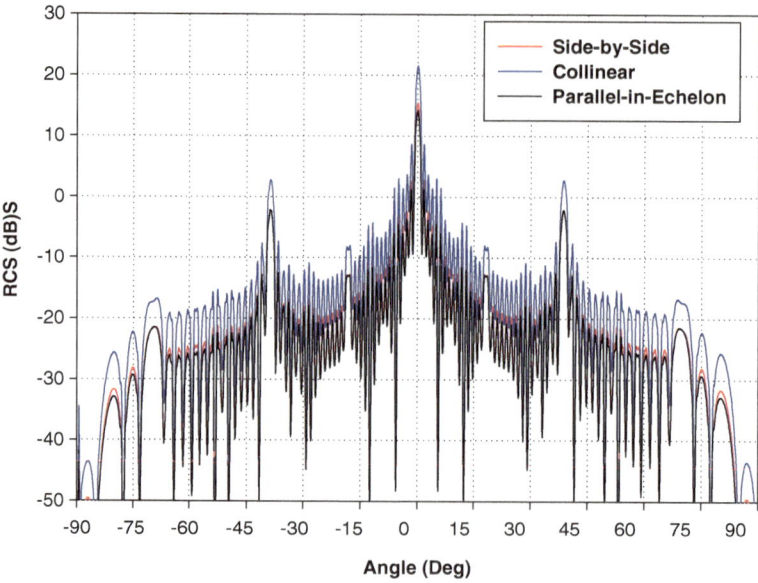

Fig. 4 Effect of array configuration on the RCS of 64-element linear parallel-fed dipole array

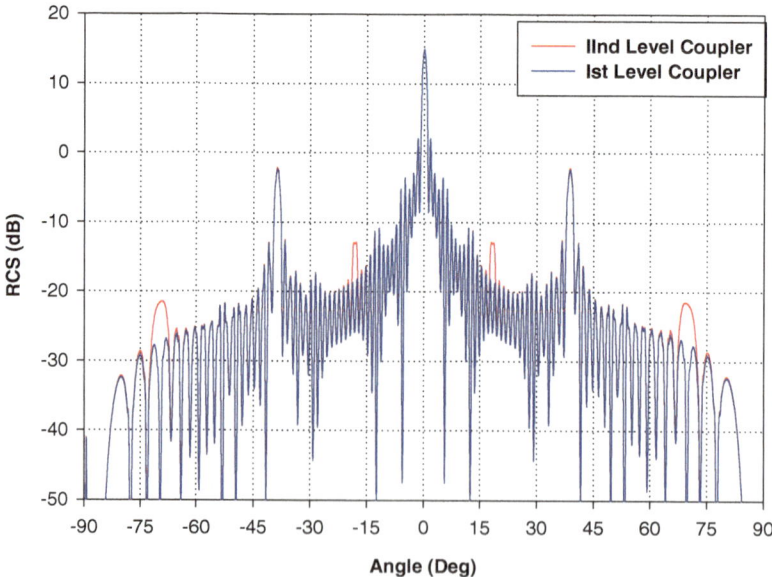

Fig. 5 Role of coupler level in RCS pattern of 64-element linear side-by-side parallel-fed dipole array

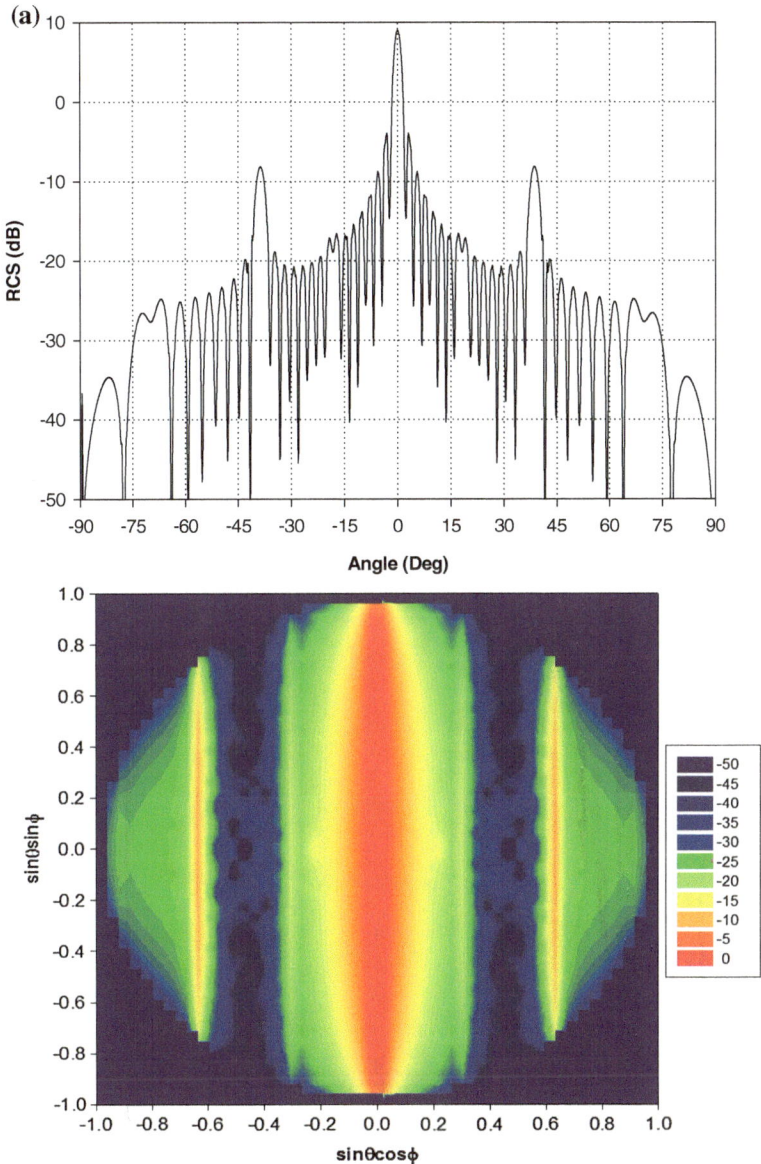

Fig. 6 Effect of beam scanning on RCS pattern of 32-element linear side-by-side dipole array. **a** $\theta_s = 0°$. **b** $\theta_s = 30°$. **c** $\theta_s = 60°$

RCS value of the lobe towards the scan angle increases. Moreover with increase in beam scan angle, extra lobes arise in the RCS pattern, but with decrease in the specular lobe RCS.

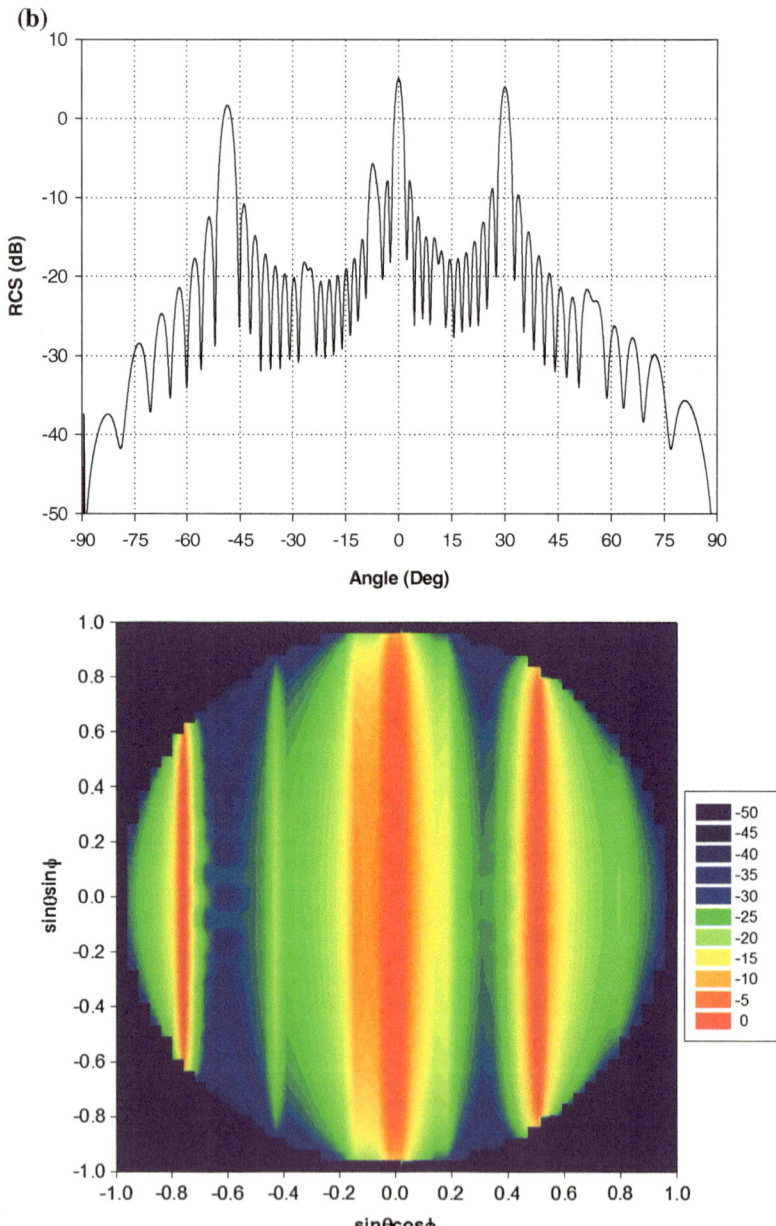

Fig. 6 (continued)

Next, effect of terminating load impedance on RCS pattern of 64-element parallel-in-echelon dipole array is shown in Fig. 7.

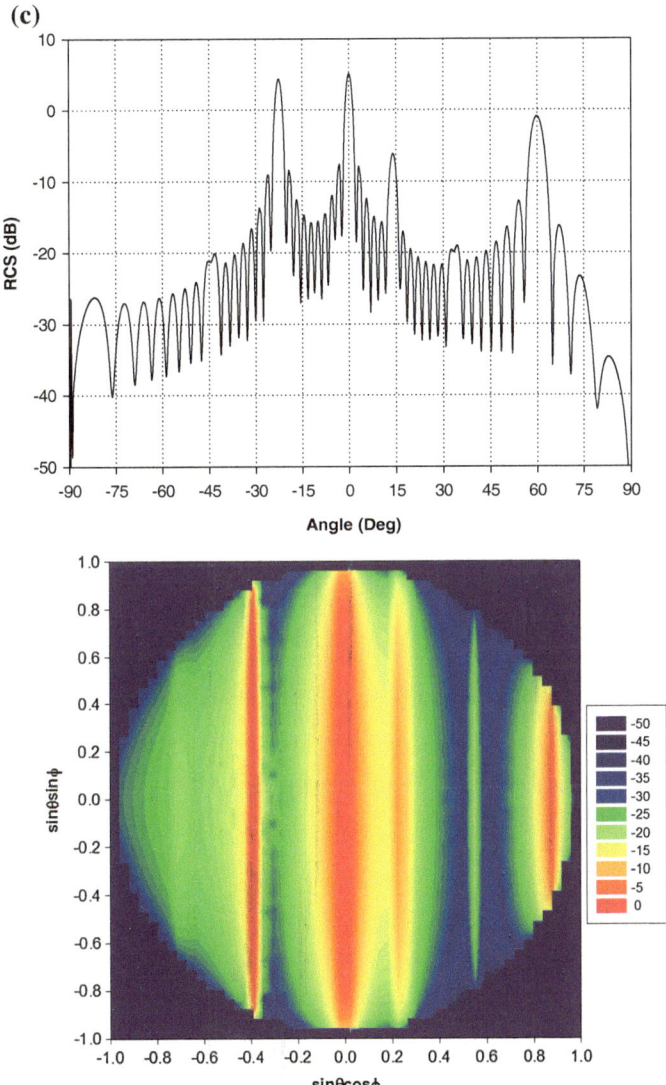

Fig. 6 (continued)

   As the impedance value of terminating load is increased from 50 to 90 Ω, the specular lobe RCS decreases. However on further increase in the value of terminating load, i.e. 120 and 180 Ω, the specular lobe RCS increases. This shows that there is limiting value of terminating load impedance towards low specular RCS of dipole array. However the trend of variation in Bragg's lobes and the lobes due to coupler mismatches are different from that of specular lobe. It may be seen that the RCS value at every lobe is lowest for terminating load impedance of 90 Ω.

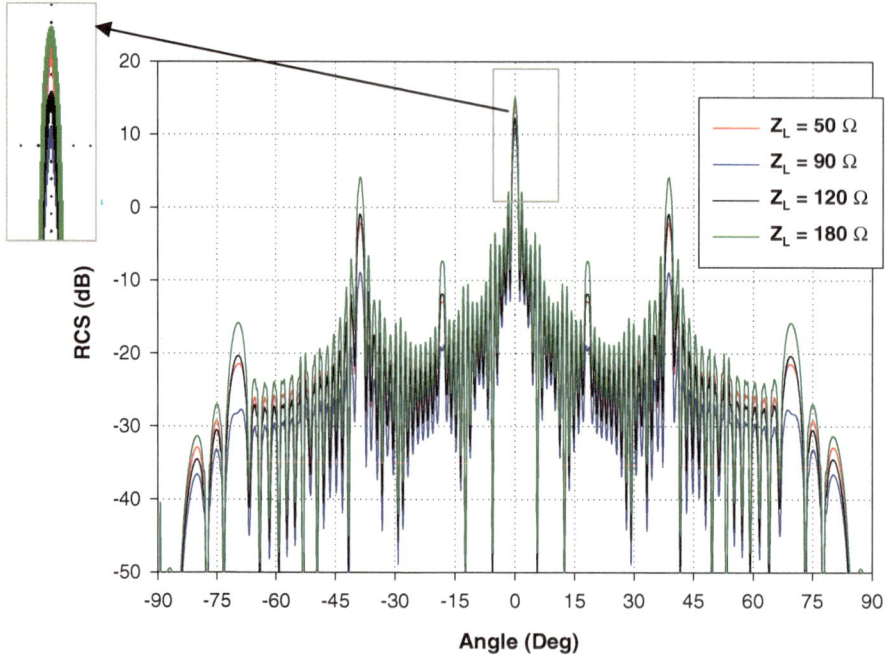

Fig. 7 Dependence of array RCS on terminating load impedance for a 64-element linear parallel-in-echelon dipole array

4 RCS of Parallel-Fed Planar Dipole Array Using Approximate Model

Next, the RCS of a parallel-fed planar dipole array is considered. A planar array is modeled as stack of linear dipole array along x-direction (Fig. 8). Here the half-wavelength dipoles are placed along both x and y directions. Similar to (2) of linear dipole array, the scattered field of mnth antenna element in planar array, $\vec{E}^s_{mn}$ is expressed as

$$\vec{E}^s_{mn}(\theta, \phi) = \left[\frac{j\eta}{4\lambda Z_{a_{mn}}} \vec{h}\left\{ \vec{h} \cdot \vec{E}^i(\theta, \phi) \right\} \frac{e^{-j\vec{k}\vec{R}}}{R} \right] \Gamma^r_{mn}(\theta, \phi) \tag{31}$$

$$\text{or } \vec{E}^s_{mn}(\theta, \phi) = 4\pi \left[\sum_{m=1}^{N_x} \sum_{n=1}^{N_y} \left\{ \frac{j\eta_o}{4\lambda Z_{a_{mn}}} \left(\int_{\Delta l} \cos(kl) \mathrm{d}l \right)^2 \cdot \cos\theta \cdot \vec{E}^r_{mn}(\theta, \phi) \right\} \right] \tag{32}$$

Here, $\vec{E}^r_{mn}(\theta, \phi) = \Gamma^r_{mn}(\theta, \phi)\, e^{j(m-1)\alpha + (n-1)\beta}$, $\beta = kd_y \sin\theta \sin\phi d_y$ is the inter-element spacing along the y-direction.

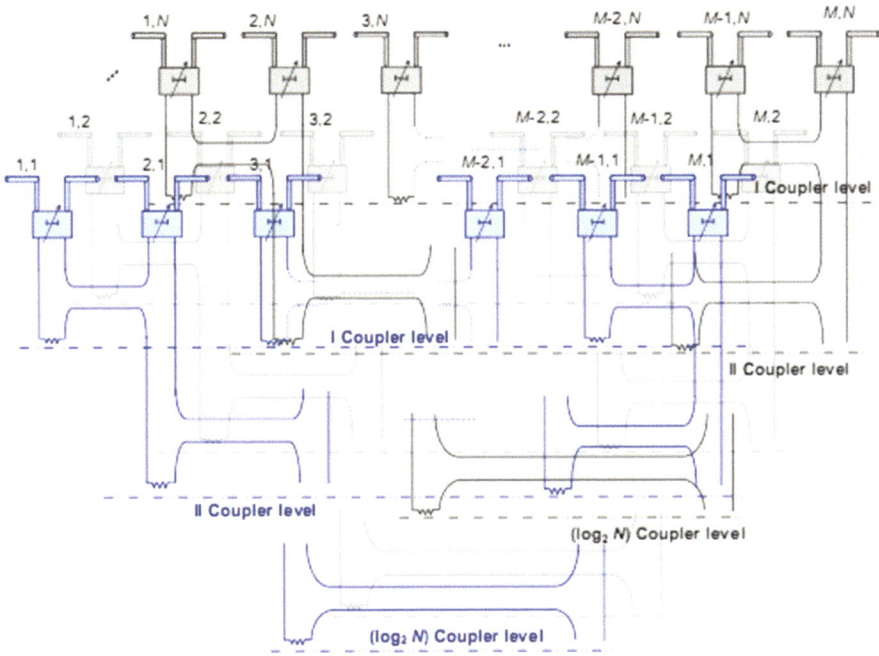

Fig. 8 Typical parallel feed network of planar dipole array; $M = N_x$, $N = N_y$

The array RCS is expressed as

$$\sigma(\theta, \phi) = 4\pi \left| F \sum_{m=1}^{N_x} \sum_{n=1}^{N_y} \vec{E}_{mn}^r(\theta, \phi) \right|^2 \tag{33}$$

The reflected field $\vec{E}_{mn}^r(\theta, \phi)$ consists of the fields scattered for different components due to impedance mismatches within the feed network. The impedance $Z_{a_{mn}}$ in (32) includes the mutual impedance. According to Kirchoff's law, the induced voltage at mnth element due to current flowing in pqth element is expressed as

$$V_{mn} = \sum_{p=1}^{N_x} \sum_{q=1}^{N_y} Z_{mn,pq} I_{pq} \tag{34}$$

where V_{mn} defines the terminal voltage at mnth element due to unity current flowing in antenna element pqth element ($mn \neq pq$) when the current in all the other elements is zero. Thus the $Z_{mn,pq}$ terms represent the mutual impedances when ($mn \neq pq$), otherwise it will be called as self impedance.

The aperture distribution plays a vital role for in feed current of the antenna element. Here, three dipole array configurations (Table 1) are considered. These are

Table 1 Dipole array configurations

Condition	Configuration	Parameters
$m = p$	Side-by-side	$d_r = |q - n|d_x, l_n$
$n = q$	Collinear	$d_r = |p - m|d_y, l_n$
$m \neq p; n \neq q$	Parallel-in-echelon	$d_r = |q - n|d_x, h_r = |p - m|d_y, l_n$

side-by-side, collinear and parallel-in-echelon. The driving impedance of mnth dipole element is defined as

$$Z_{a_{mn}} = \frac{V_{mn}}{I_{mn}} = \sum_{p=1}^{N_x} \sum_{q=1}^{N_y} Z_{mn,pq} \frac{I_{pq}}{I_{mn}} \tag{35}$$

The antenna impedance is expressed as $Z_{a_{mn}} = R_{a_{mn}} + jX_{a_{mn}}$. The self impedance of a dipole is given by (Balanis 2005)

$$R_{\text{self}_n} = \frac{\eta}{2\pi} \left[\begin{array}{c} C + \ln(kl_n) - C_i(kl_n) + \frac{1}{2}\sin(kl_n)\{S_i(2kl_n) - 2S_i(kl_n)\} \\ + \frac{1}{2}\cos(kl_n)\left\{ C + \ln\left(kl_n/2\right) + C_i(2kl_n) - 2C_i(kl_n) \right\} \end{array} \right] \tag{36a}$$

$$X_{\text{self}_n} = \frac{\eta}{4\pi} \left[\begin{array}{c} 2S_i(kl_n) + \cos(kl_n)\{2S_i(kl_n) - S_i(2kl_n)\} \\ - \sin(kl_n)\left\{ 2C_i(kl_n) - C_i(2kl_n) - C_i\left(\frac{2ka_n^2}{l_n}\right) \right\} \end{array} \right] \tag{36b}$$

where $C_i(kl_n)$ and $S_i(kl_n)$ are cosine and sine integrals expressed as

$$S_i(x) = \sum_{k=0}^{\infty} \frac{(-1)^k x^{2k+1}}{(2k+1)(2k+1)!} \quad C_i(x) = C + \ln(x) + \sum_{k=1}^{\infty} (-1)^k \frac{x^{2k}}{2k(2k)!}$$

The expressions of the mutual impedances for different configurations are given below. The parameters d_r, h_r, and l_n, are used according to the configuration of dipole pair in the planar dipole array, given in Table 1.

Side-by-Side Configuration

The dipole elements are arranged uniformly in side-by-side configuration along x-axis.

$$R_{s_r} = \frac{\eta}{4\pi} \left[2C_i(kd_r) - C_i\left(k\left(\sqrt{d_r^2 + l_n^2} + l_n\right)\right) - C_i\left(k\left(\sqrt{d_r^2 + l_n^2} - l_n\right)\right) \right] \tag{36c}$$

$$X_{s_r} = -\frac{\eta}{4\pi} \left[2S_i(kd_r) - S_i\left(k\left(\sqrt{d_r^2 + l_n^2} + l_n\right)\right) - S_i\left(k\left(\sqrt{d_r^2 + l_n^2} - l_n\right)\right) \right] \tag{36d}$$

Collinear Configuration

The dipole elements are arranged uniformly in collinearly along y-axis.

$$
R_{c_r} = -\frac{\eta}{8\pi}\cos(k(d_r + l_n))\begin{bmatrix} -2C_i(2k(d_r + l_n)) + C_i(2kd_r) \\ +C_i(2k(d_r + 2l_n)) \\ -\ln\left(\left(d_r^2 + 2d_r l_n\right)\big/(d_r + l_n)^2\right) \end{bmatrix}
$$
$$
+\frac{\eta}{8\pi}\sin(k(d_r + l_n))\begin{bmatrix} 2S_i(2k(d_r + l_n)) - S_i(2kd_r) \\ -S_i(2k(d_r + 2l_n)) \end{bmatrix} \tag{36e}
$$

$$
X_{c_r} = -\frac{\eta}{8\pi}\cos(k(d_r + l_n))\begin{bmatrix} 2S_i(2k(d_r + l_n)) - S_i(2kd_r) \\ -S_i(2k(d_r + 2l_n)) \end{bmatrix}
$$
$$
+\frac{\eta}{8\pi}\sin(k(d_r + l_n))\begin{bmatrix} 2C_i(2k(d_r + l_n)) - C_i(2kd_r) \\ -C_i(2k(d_r + 2l_n)) \\ -\ln\left(\left(d_r^2 + 2d_r l_n\right)\big/(d_r + l_n)^2\right) \end{bmatrix} \tag{36f}
$$

Parallel-in-Echelon Configuration

The dipole elements are arranged in such a way that each second dipole element is at height h with reference to the feed point of centre-fed first dipole element.

$$
R_{p_r} = -\frac{\eta}{8\pi}\cos(kh_r)\begin{bmatrix} -2C_i\left(k\left(\sqrt{d_r^2 + h_r^2} + h_r\right)\right) \\ -2C_i\left(k\left(\sqrt{d_r^2 + h_r^2} - h_r\right)\right) \\ +C_i\left(k\left(\sqrt{d_r^2 + (h_r - l_n)^2} + (h_r - l_n)\right)\right) \\ +C_i\left(k\left(\sqrt{d_r^2 + (h_r - l_n)^2} - (h_r - l_n)\right)\right) \\ +C_i\left(k\left(\sqrt{d_r^2 + (h_r + l_n)^2} + (h_r + l_n)\right)\right) \\ +C_i\left(k\left(\sqrt{d_r^2 + (h_r + l_n)^2} - (h_r + l_n)\right)\right) \end{bmatrix}
$$
$$
+\frac{\eta}{8\pi}\sin(kh_r)\begin{bmatrix} 2S_i\left(k\left(\sqrt{d_r^2 + h_r^2} + h_r\right)\right) - 2S_i\left(k\left(\sqrt{d_r^2 + h_r^2} - h_r\right)\right) \\ -S_i\left(k\left(\sqrt{d_r^2 + (h_r - l_n)^2} + (h_r - l_n)\right)\right) \\ +S_i\left(k\left(\sqrt{d_r^2 + (h_r - l_n)^2} - (h_r - l_n)\right)\right) \\ -S_i\left(k\left(\sqrt{d_r^2 + (h_r + l_n)^2} + (h_r + l_n)\right)\right) \\ +S_i\left(k\left(\sqrt{d_r^2 + (h_r + l_n)^2} - (h_r + l_n)\right)\right) \end{bmatrix} \tag{36g}
$$

$$
X_{p_r} = -\frac{\eta}{8\pi}\cos(kh_r)
\begin{bmatrix}
2S_i\big(k\big(\sqrt{d_r^2 + h_r^2} + h_r\big)\big) + 2S_i\big(k\big(\sqrt{d_r^2 + h_r^2} - h_r\big)\big) \\
- S_i\left(k\left(\sqrt{d_r^2 + (h_r - l_n)^2} + (h_r - l_n)\right)\right) \\
- S_i\left(k\left(\sqrt{d_r^2 + (h_r - l_n)^2} - (h_r - l_n)\right)\right) \\
- S_i\left(k\left(\sqrt{d_r^2 + (h_r + l_n)^2} + (h_r + l_n)\right)\right) \\
- S_i\left(k\left(\sqrt{d_r^2 + (h_r + l_n)^2} - (h_r + l_n)\right)\right)
\end{bmatrix}
$$

$$
+\frac{\eta}{8\pi}\sin(kh_r)
\begin{bmatrix}
2C_i\big(k\big(\sqrt{d_r^2 + h_r^2} + h_r\big)\big) - 2C_i\big(k\big(\sqrt{d_r^2 + h_r^2} - h_r\big)\big) \\
- C_i\left(k\left(\sqrt{d_r^2 + (h_r - l_n)^2} + (h_r - l_n)\right)\right) \\
+ C_i\left(k\left(\sqrt{d_r^2 + (h_r - l_n)^2} - (h_r - l_n)\right)\right) \\
- C_i\left(k\left(\sqrt{d_r^2 + (h_r + l_n)^2} + (h_r + l_n)\right)\right) \\
+ C_i\left(k\left(\sqrt{d_r^2 + (h_r + l_n)^2} - (h_r + l_n)\right)\right)
\end{bmatrix}
$$

$$(36h)$$

Radiators The RCS estimation of radiators in planar dipole array with parallel feed network is carried out as in the case of linear dipole array. The signal path is followed as it enters into the array aperture and travels through the feed network, the RCS contribution by the radiators for $N_x \times N_y$ dipole array is given by

$$
\sigma_r(\theta, \phi) = \frac{4\pi}{\lambda^2}\left| F \sum_{m=1}^{N_x}\sum_{n=1}^{N_y} \vec{E}_{r_{mn}}^r(\theta, \phi)\right|^2 = \frac{4\pi}{\lambda^2}\left| F \sum_{m=1}^{N_x}\sum_{n=1}^{N_y} \Gamma_{r_{mn}} e^{2j\{(m-1)\alpha + (n-1)\beta\}}\right|^2
$$

$$
= \frac{4\pi}{\lambda^2}\left| F \left[\sum_{m=1}^{N_x}\sum_{n=1}^{N_y} \Gamma_{r_{mn}}\right]\left[\sum_{n=1}^{N_y} e^{2j(n-1)\beta}\right]\left[\sum_{m=1}^{N_x} e^{2j(m-1)\alpha}\right]\right|^2
$$

$$(37)$$

$$
\sigma_r(\theta, \phi) = \frac{4\pi}{\lambda^2}\left| F \left[\sum_{m=1}^{N_x}\sum_{n=1}^{N_y} \Gamma_{r_{mn}}\right]\left[\sum_{q=0}^{N_y-1} e^{2jq\beta}\right]\left[\sum_{p=0}^{N_x-1} e^{2jp\alpha}\right]\right|^2 \quad ; m-1 = p; n-1 = q
$$

$$(37a)$$

In closed form, the expression (37a) is written as

$$\sigma_r(\theta, \phi)|_{\text{normalized}} = \frac{4\pi}{\lambda^2} \left| F \cdot \frac{\sin(\alpha N_x)}{N_x \sin \alpha} \cdot \frac{\sin(\beta N_y)}{N_y \sin \beta} \cdot \sum_{m=1}^{N_x} \sum_{n=1}^{N_y} \Gamma_{r_{mn}} \right|^2 \tag{37b}$$

where, $\Gamma_{r_{mn}} = \left| \frac{Z_{a_{mn}} - Z_o}{Z_{a_{mn}} + Z_o} \right|$ is the reflection coefficient of the radiator.

Phase shifters The radiators are connected to the phase-shifters in a feed network. Here, the phase shifters are modeled as simple lossless delay lines with characteristic impedance Z_o. The length of the delay lines are taken on the base of phase-shift required. The length of the delay-lines, L_{mn} is obtained from the phase-shift as

$$L_{mn} = \frac{\lambda}{2\pi} \left[(m-1)kd_x \sin \theta_s \cos \phi_s + (n-1)kd_y \sin \theta_s \sin \phi_s \right] \tag{38}$$

These phase shifters are connected to the antenna terminals at one end and to the couplers on the other end. The antenna impedance, which acts as an input impedance for the phase shifter, will be translated along the length of delay line. This yields the impedance at the other end of the phase shifter, i.e. $Z_{p_{mn}}$, expressed as

$$Z_{p_{mn}} = Z_o \left[\frac{Z_{a_{mn}} + jZ_o \tan\left(\frac{2\pi}{\lambda} L_{mn}\right)}{Z_o + jZ_{a_{mn}} \tan\left(\frac{2\pi}{\lambda} L_{mn}\right)} \right] \tag{39}$$

The corresponding RCS of phase shifters is given by

$$\sigma_p(\theta, \phi) = \frac{4\pi}{\lambda^2} \left| F \sum_{m=1}^{N_x} \sum_{n=1}^{N_y} \vec{E}_{p_{mn}}^r (\theta, \phi) \right|^2$$

$$= \frac{4\pi}{\lambda^2} \left| F \sum_{m=1}^{N_x} \sum_{n=1}^{N_y} \Gamma_{p_{mn}} T_{r_{mn}}^2 e^{2j\{(m-1)\alpha + (n-1)\beta\}} \right|^2 \tag{40}$$

$$= \frac{4\pi}{\lambda^2} \left| F \left[\sum_{m=1}^{N_x} \sum_{n=1}^{N_y} \Gamma_{p_{mn}} \Gamma_{r_{mn}}^2 \right] \left[\sum_{n=1}^{N_y} e^{2j(n-1)\beta} \right] \left[\sum_{m=1}^{N_x} e^{2j(m-1)\alpha} \right] \right|^2$$

$$\text{or } \sigma_p(\theta, \phi)|_{\text{normalized}} = \frac{4\pi}{\lambda^2} \left| F \cdot \frac{\sin(\alpha N_x)}{N_x \sin \alpha} \cdot \frac{\sin(\beta N_y)}{N_y \sin \beta} \cdot \sum_{m=1}^{N_x} \sum_{n=1}^{N_y} \Gamma_{p_{mn}} T_{r_{mn}}^2 \right|^2 \tag{40a}$$

$\Gamma_{p_{mn}} = \left| \frac{Z_{p_{mn}} - Z_o}{Z_{p_{mn}} + Z_o} \right|$ is the reflection coefficient of the phase shifter, and $|T_{r_{mn}}|^2 = 1 - |\Gamma_{r_{mn}}|^2$ is the transmission coefficient of the radiators.

Couplers The couplers come after phase shifters in a parallel feed. The RCS obtained using approximate model is given by

$$\sigma_{cp}(\theta, \phi)\big|_{\text{normalized}} = \frac{4\pi}{\lambda^2} \left| F \sum_{m=1}^{N_x} \sum_{n=1}^{N_y} \vec{E}^r_{cp_{mn}}(\theta, \phi) \right|^2$$

$$= \frac{4\pi}{\lambda^2} \left| F \frac{\sin(\xi_x N_x)}{N_x \sin \xi_x} \frac{\sin(\xi_y N_y)}{N_y \sin \xi_y} \sum_{m=1}^{N_x} \sum_{n=1}^{N_y} T^2_{r_{mn}} T^2_{p_{mn}} \Gamma_{cp_{mn}} \right|^2 \tag{41}$$

where,

$$\Gamma_{cp_{mn}} = \left| \frac{Z_{33_{1i}} - Z_{p_{mn}}}{Z_{33_{1i}} + Z_{p_{mn}}} \right| \text{ for odd-numbered elements, i.e., at Port 3} \tag{41a}$$

$$\Gamma_{cp_{mn}} = \left| \frac{Z_{22_{1i}} - Z_{p_{mn}}}{Z_{22_{1i}} + Z_{p_{mn}}} \right| \text{ for even-numbered elements, i.e., at Port 2} \tag{41b}$$

where $T_{p_{mn}}$ is the transmission coefficient of *mn*th phase-shifter, $\Gamma_{cp_{mn}}$ is the reflection coefficient of *mn*th coupler, $\zeta_x = \alpha + \alpha_s$; α_s is the inter-element phase to scan antenna beam in *x*-direction, and $\zeta_y = \beta + \beta_s$; with β_s as inter-element phase to scan antenna beam in *y*-direction.

Sum and difference arms of first level couplers The signal at the input ports of coupler travels towards Port 1 and/or Port 4 of the first level couplers. Thus the reflected signals become significant at the sum and difference arms of first level couplers. The corresponding reflection coefficients are $\Gamma_{s_{qi}}$ and $\Gamma_{d_{qi}}$ for sum and difference ports, respectively, where sub-suffices *q* and *i* represents the coupler-level and the coupler number in order. The reflection coefficients at the sum and difference arms of the couplers for different input ports are expressed as (Sneha et al. 2013c)

$$\Gamma_{s_{qi}} = \left| \frac{Z_{31_{qi}} - Z_{11_{qi}}}{Z_{31_{qi}} + Z_{11_{qi}}} \right| \quad \text{for odd-arms of the coupler} \tag{42}$$

$$\Gamma_{s_{qi}} = \left| \frac{Z_{21_{qi}} - Z_{11_{qi}}}{Z_{21_{qi}} + Z_{11_{qi}}} \right| \quad \text{for even-arms of the coupler} \tag{42a}$$

$$\Gamma_{d_{qi}} = \left| \frac{Z_{34_{qi}} - Z_{44_{qi}}}{Z_{34_{qi}} + Z_{44_{qi}}} \right| \quad \text{for odd-arms of the coupler} \tag{42b}$$

$$\Gamma_{d_{qi}} = \left| \frac{Z_{24_{qi}} - Z_{44_{qi}}}{Z_{24_{qi}} + Z_{44_{qi}}} \right| \quad \text{for even-arms of the coupler} \tag{42c}$$

The transmission coefficient at the sum port of couplers is given by

$$T_{s_{qi}} = \sqrt{1 - \Gamma_{s_{qi}}^2}$$ (43)

Due to the impedance mismatches at the sum and difference arm of first level couplers, the RCS at mnth dipole element for a given n, is given by

$$\sigma_{sd_1}(\theta, \phi)_{mn}|_{\text{normalized}} = \frac{4\pi}{\lambda^2}\left|F\frac{\sin(N_x\xi_x)}{N_x\sin(2\xi_x)}\frac{\sin(N_y\xi_y)}{N_y\sin\xi_y}\right.$$

$$\left.\times\sum_{n=1}^{N_y}\sum_{m=1,3\ldots}^{N_x-1}T_{r_{mn}}T_{p_{mn}}\left\{c_{1i}e^{j\psi}\Gamma_{s_{1i}}\begin{pmatrix}c_{1i}e^{j\psi}T_{r_{mn}}T_{p_{mn}}\\+T_{r_{(m+1)n}}T_{p_{(m+1)n}}T_{c_{1i}}\end{pmatrix}\\+\Gamma_{d_{1i}}T_{c_{1i}}\begin{pmatrix}T_{c_{1i}}T_{r_{mn}}T_{p_{mn}}\\+T_{r_{(m+1)n}}T_{p_{(m+1)n}}c_{1i}e^{j\psi}\end{pmatrix}\right\}\right|^2$$ (44)

Similarly, RCS at $(m + 1)$th element for a given n due to sum and difference port of first level couplers is given by

$$\sigma_{sd_1}(\theta, \phi)_{(m+1)n}|_{\text{normalized}} = \frac{4\pi}{\lambda^2}\left|F\frac{\sin(N_x\xi_x)}{N_x\sin(2\xi_x)}\frac{\sin(N_y\xi_y)}{N_y\sin\xi_y}\sum_{n=1}^{N_y}\sum_{m=1,3\ldots}^{N_x-1}T_{r_{(m+1)n}}T_{p_{(m+1)n}}\right.$$

$$\left.\times\left\{\Gamma_{s_{1i}}T_{c_{1i}}\begin{pmatrix}T_{r_{mn}}T_{p_{mn}}c_{1i}e^{j\psi}\\+T_{c_{1i}}T_{r_{(m+1)n}}T_{p_{(m+1)n}}\end{pmatrix}\\+\Gamma_{d_{1i}}e^{j\psi}c_{1i}\begin{pmatrix}T_{r_{mn}}T_{p_{mn}}T_{c_{1i}}\\+T_{r_{(m+1)n}}T_{p_{(m+1)n}}c_{1i}e^{j\psi}\end{pmatrix}\right\}\right|^2$$ (45)

In a first level coupler, two adjacent dipole elements are connected to a single coupler. Thus the total reflected field at the first level coupler level is the sum of the fields reflected at two dipole elements. The corresponding RCS equation is

$$\sigma_{sd_1}(\theta, \phi)|_{\text{normalized}} = \left[\sigma_{sd_1}(\theta, \phi)_{mn}|_{\text{normalized}} + \sigma_{sd_1}(\theta, \phi)_{(m+1)n}|_{\text{normalized}}\right]$$ (46)

Adding (37b), (40a), (41) and (46), the total RCS of the planar dipole array due to the mismatches in the feed network till first level of couplers is obtained as

$$\sigma(\theta, \phi)|_{\text{normalized}} = \left[\sigma_r(\theta, \phi)|_{\text{normalized}}\right] + \left[\sigma_p(\theta, \phi)|_{\text{normalized}}\right]$$
$$+ \left[\sigma_{cp}(\theta, \phi)|_{\text{normalized}}\right] + \left[\sigma_{sd_1}(\theta, \phi)|_{\text{normalized}}\right]$$ (47)

Sum and difference arms of second level couplers In a second level of couplers in a parallel feed, four adjacent dipole elements are connected to a single coupler. This implies that the reflected signal at *mn*th array element (for $m = 1, 5, 9..., n$) due to the reflections from second level coupler comprises of the signals from mn, $(m + 1)n$, $(m + 2)n$ and $(m + 3)n$ elements.

The RCS at *mn*th dipole element due to the reflections till second level couplers using approximate method is given by

$$
\sigma_{\mathrm{sd2}}(\theta, \phi)_{mn}\big|_{\mathrm{normalized}} = \frac{4\pi}{\lambda^2} \left| F \frac{\sin(N_y \xi_y)}{N_y \sin \xi_y} \frac{\sin(N_x \xi_x)}{N_x \sin(4\xi_y)} \sum_{n=1}^{N_y} \sum_{m=1}^{N_x-3} T_{s_{li}} c_{li} e^{j\psi} T_{r_{mn}} T_{p_{mn}} \right.
$$

$$
\times \left[\begin{array}{l} \Gamma_{s2i'} c_{2i'} e^{j\psi} \left\{ \begin{array}{l} T_{r_{mn}} T_{p_{mn}} c_{li} e^{j\psi} T_{s_{li}} c_{2i'} e^{j\psi} + T_{r_{(m+1)n}} T_{p_{(m+1)n}} T_{c_{li}} T_{s_{li}} c_{2i'} e^{j\psi} \\ + T_{r_{(m+2)n}} T_{p_{(m+2)n}} c_{1(i+1)} e^{j\psi} T_{s_{1(i+1)}} T_{c_{2i'}} \\ + T_{r_{(m+3)n}} T_{p_{(m+3)n}} T_{c_{1(i+1)}} T_{s_{1(i+1)}} T_{c_{2i'}} \end{array} \right\} \\ + \Gamma_{d2i'} T_{c_{2i'}} \left\{ \begin{array}{l} T_{r_{mn}} T_{p_{mn}} c_{li} e^{j\psi} T_{s_{li}} T_{c_{2i'}} + T_{r_{(m+1)n}} T_{p_{(m+1)n}} T_{c_{li}} T_{s_{li}} T_{c_{2i'}} \\ + T_{r_{(m+2)n}} T_{p_{(m+2)n}} c_{1(i+1)} e^{j\psi} T_{s_{1(i+1)}} \\ c_{2i'} e^{j\psi} + T_{r_{(m+3)n}} T_{p_{(m+3)n}} T_{c_{1(i+1)}} T_{s_{1(i+1)}} c_{2i'} e^{j\psi} \end{array} \right\} \end{array} \right]^2
$$

$$(48)$$

Next, the reflected field at $(m + 1)$th dipole for given n, due to the reflections at sum and difference ports of second level coupler is given by

$$
\sigma_{\mathrm{sd2}}(\theta, \phi)_{(m+1)n}\big|_{\mathrm{normalized}} = \frac{4\pi}{\lambda^2} \left| \begin{array}{l} F \frac{\sin(N_y \xi_y)}{N_y \sin \xi_y} \frac{\sin(N_x \xi_x)}{N_x \sin(4\xi_x)} \\ \times \sum_{n=1}^{N_y} \sum_{m=1}^{N_x-3} T_{r_{(m+1)n}} T_{p_{(m+1)n}} T_{c_{li}} T_{s_{li}} \end{array} \right.
$$

$$
\times \left[\begin{array}{l} \Gamma_{s2i'} c_{2i'} e^{j\psi} \left\{ \begin{array}{l} T_{r_{mn}} T_{p_{mn}} c_{1i} e^{j\psi} T_{s_{1i}} c_{2i'} e^{j\psi} \\ + T_{r_{(m+1)n}} T_{p_{(m+1)n}} T_{c_{1i}} T_{s_{1i}} c_{2i'} e^{j\psi} \\ + T_{r_{(m+2)n}} T_{p_{(m+2)n}} c_{1(i+1)} \\ e^{j\psi} T_{s_{1(i+1)}} T_{c_{2i'}} + T_{r_{(m+3)n}} T_{p_{(m+3)n}} \\ T_{c_{1(i+1)}} T_{s_{1(i+1)}} T_{c_{2i'}} \end{array} \right\} \\ + \Gamma_{d2i'} T_{c_{2i'}} \left\{ \begin{array}{l} T_{r_{mn}} T_{p_{mn}} c_{1i} e^{j\psi} T_{s_{1i}} T_{c_{2i'}} \\ + T_{r_{(m+1)n}} T_{p_{(m+1)n}} T_{c_{1i}} T_{s_{1i}} T_{c_{2i'}} \\ + T_{r_{(m+2)n}} T_{p_{(m+2)n}} c_{1(i+1)} \\ e^{j\psi} T_{s_{1(i+1)}} c_{2i'} e^{j\psi} + T_{r_{(m+3)n}} \\ T_{p_{(m+3)n}} T_{c_{1(i+1)}} T_{s_{1(i+1)}} c_{2i'} e^{j\psi} \end{array} \right\} \end{array} \right]^2
$$

$$(49)$$

Likewise, the signal reflections at $(m + 2)$th and $(m + 3)$th dipole ($m = 1, 5, 9...$) for given n due to the reflections at sum and difference ports of second level coupler, are expressed as

$$
\sigma_{\mathrm{sd}_2}(\theta,\phi)_{(m+2)n}\big|_{\mathrm{normalized}} = \frac{4\pi}{\lambda^2}\left| F\, \frac{\sin(N_y\xi_y)}{N_y\sin\xi_y}\frac{\sin(N_x\xi_x)}{N_x\sin(4\xi_x)} \right.
$$

$$
\times \sum_{n=1}^{N_y}\sum_{m=1}^{Nx-3} T_{r(m+2)n}T_{p(m+2)n}c_{1(i+1)}e^{j\psi}T_{s_{1(i+1)}}
$$

$$
\left[\Gamma_{s_{2i'}}T_{c_{2i'}}\left\{ \begin{array}{l} T_{r_{mn}}T_{p_{mn}}c_{1i}e^{j\psi}T_{s_{1i}} \\ c_{2i'}e^{j\psi}+T_{r(m+1)n}T_{p(m+1)n} \\ T_{c_{1i}}T_{s_{1i}}c_{2i'}e^{j\psi} \\ +T_{r(m+2)n}T_{p(m+2)n} \\ e^{j\psi}c_{1(i+1)}T_{s_{1(i+1)}} \\ T_{c_{2i'}}+T_{r(m+3)n}T_{p(m+3)n} \\ T_{c_{1(i+1)}}T_{s_{1(i+1)}}T_{c_{2i'}} \end{array}\right\} + \Gamma_{d_{2i'}}c_{2i'}e^{j\psi}\left\{ \begin{array}{l} T_{r_{mn}}T_{p_{mn}}c_{1i}e^{j\psi}T_{s_{1i}} \\ T_{c_{2i'}}+T_{r(m+1)n}T_{p(m+1)n} \\ T_{c_{1i}}T_{s_{1i}}T_{c_{2i'}} \\ +T_{r(m+2)n}T_{p(m+2)n} \\ e^{j\psi}c_{1(i+1)}T_{s_{1(i+1)}} \\ \times c_{2i'}e^{j\psi} \\ +T_{r(m+3)}T_{p(m+3)}T_{c_{1(i+1)}} \\ \times T_{s_{1(i+1)}}c_{2i'}e^{j\psi} \end{array}\right\} \right]\Bigg|^{\,2}
$$

$$(50)$$

$$
\sigma_{\mathrm{sd}_2}(\theta,\phi)_{(m+3)n}\big|_{\mathrm{normalized}} = \frac{4\pi}{\lambda^2}\left| F\, \frac{\sin(N_y\xi_y)}{N_y\sin\xi_y}\frac{\sin(N_x\xi_x)}{N_x\sin(4\xi_x)} \right.
$$

$$
\times \sum_{n=1}^{N_y}\sum_{m=1}^{Nx-3} T_{r(m+3)n}T_{p(m+3)n}T_{s_{1(i+1)}}T_{c_{1(i+1)}}
$$

$$
\times \left[\Gamma_{s_{2i'}}T_{c_{2i'}}\left\{ \begin{array}{l} T_{r_{mn}}T_{p(m+3)n}c_{1i}e^{j\psi}T_{s_{1i}}c_{2i'}e^{j\psi} \\ +T_{r(m+1)n}T_{p(m+1)n}T_{c_{1i}}T_{s_{1i}}c_{2i'}e^{j\psi} \\ +T_{r(m+2)n}T_{p(m+2)n}c_{1(i+1)}e^{j\psi}T_{s_{1(i+1)}}T_{c_{2i'}} \\ +T_{r(m+3)n}T_{p(m+3)n}T_{c_{1(i+1)}}T_{s_{1(i+1)}}T_{c_{2i'}} \end{array}\right\} \right.
$$

$$
\left. +\Gamma_{d_{2i'}}c_{2i'}e^{j\psi}\left\{ \begin{array}{l} T_{r_{mn}}T_{p_{mn}}c_{1i}e^{j\psi}T_{s_{1i}}T_{c_{2i'}} \\ +T_{r(m+1)n}T_{p(m+1)n}T_{c_{1i}}T_{s_{1i}}T_{c_{2i'}} \\ +T_{r(m+2)n}T_{p(m+2)n}c_{1(i+1)}e^{j\psi}T_{s_{1(i+1)}}c_{2i'}e^{j\psi} \\ +T_{r(m+3)n}T_{p(m+3)n}T_{c_{1(i+1)}}T_{s_{1(i+1)}}c_{2i'}e^{j\psi} \end{array}\right\} \right]\Bigg|^{\,2}
$$

$$(51)$$

From the Eqs. (48) through (51), the RCS of planar dipole phased array due to the mismatches at second level of couplers is expressed as

$$
\sigma_{\mathrm{sd}_2}(\theta,\phi)\big|_{\mathrm{normalized}} = \left\{ \begin{array}{l} \sigma_{\mathrm{sd}_2}(\theta,\phi)_{mn}\big|_{\mathrm{normalized}} + \sigma_{\mathrm{sd}_2}(\theta,\phi)_{(m+1)n}\big|_{\mathrm{normalized}} \\ + \sigma_{\mathrm{sd}_2}(\theta,\phi)_{(m+2)n}\big|_{\mathrm{normalized}} + \sigma_{\mathrm{sd}_2}(\theta,\phi)_{(m+3)n}\big|_{\mathrm{normalized}} \end{array}\right\}
$$

$$(52)$$

This yields the total RCS for the planar dipole array due to mismatches at radiators, phase shifters, couplers at first and second level of feed network is given by

$$\sigma(\theta, \phi)|_{\text{normalized}} = \left[\sigma_r(\theta, \phi)|_{\text{normalized}}\right] + \left[\sigma_p(\theta, \phi)|_{\text{normalized}}\right] + \left[\sigma_{cp}(\theta, \phi)|_{\text{normalized}}\right]$$
$$+ \left[\sigma_{\text{sd}_1}(\theta, \phi)|_{\text{normalized}}\right] + \left[\sigma_{\text{sd}_2}(\theta, \phi)|_{\text{normalized}}\right]$$

$$(53)$$

This analytical formulation for RCS estimation of planar dipole array can be extended to an arbitrary level of couplers in the parallel feed network. It is noted that the sub-array size required for the RCS estimation depends on the coupler level and is given by 2^q, q is the coupler level.

5   Planar Dipole Array: Simulation Results

This section discusses the RCS pattern computed for parallel-fed planar dipole array in the presence of mutual coupling effect. A systematic step-by-step approach is used to calculate RCS pattern including the finite dimensions of dipole element along with other design parameters. The RCS pattern is calculated till second level of couplers in a parallel feed. Figure 9 shows the broadside RCS pattern of a 16×16 and 32×32 dipole arrays. The inter-element spacing is taken as 0.484λ

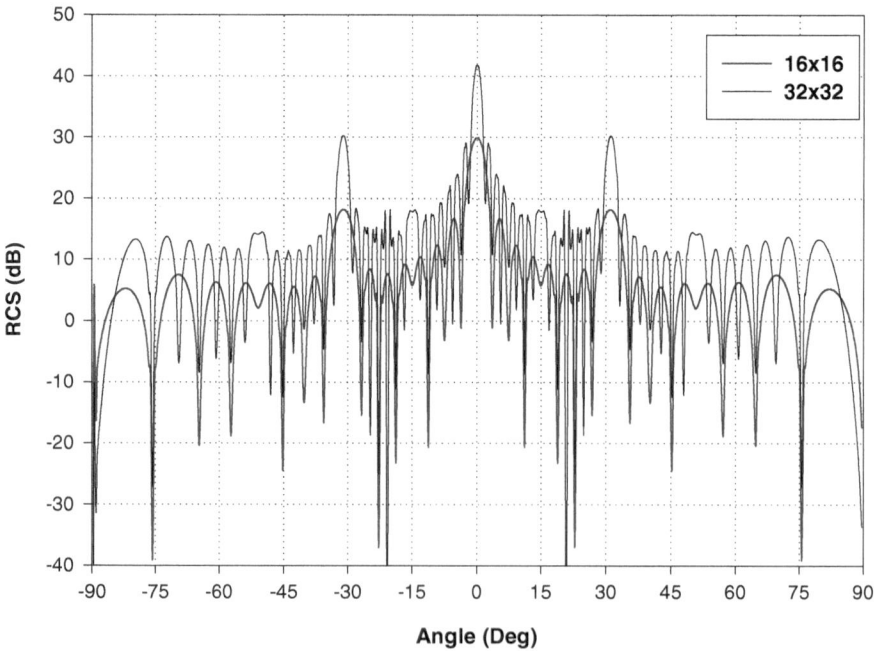

Fig. 9 Effect of array size on broadside RCS pattern of square dipole array with parallel-feed network

along x-direction and 0.77λ along y-direction. The characteristic impedance and load termination are 50 and 150 Ω respectively. The length and radius of the dipole element are taken as 0.5 and 0.001λ respectively. It may be observed from Fig. 9 that the RCS value at each point of the pattern increases with the array size and the number of elements.

The lobes due to coupler mismatches at both first and second level couplers become more prominent for larger dipole array (32×32). Another example is shown in Fig. 10. It may be noticed that RCS pattern of rectangular dipole array changes when the number of dipole elements along x-axis are increased from 16 to 20, keeping inter-element spacing constant. The spacing between dipole antenna elements is taken as 0.484λ along x-direction and 0.5λ along y-direction. The characteristic impedance and load termination are 50 and 90 Ω respectively. The level of specular lobe and lobes due to coupler mismatch is increased, with redistributed sidelobes. This is in line of expectations. The level of these lobes depends on the array size.

Next, a 16×10 dipole array is considered with inter-element spacing of 0.484 and 0.77λ along x- and y-directions respectively. Here the scan angle is varied (0°, 45°, 60°) and the RCS pattern is compared (Fig. 11). It can be seen that the specular lobe level varies with the scan angle along with emerging of additional lobes. The role of terminating load impedance in the RCS pattern of dipole array is significant (Sneha et al. 2013b).

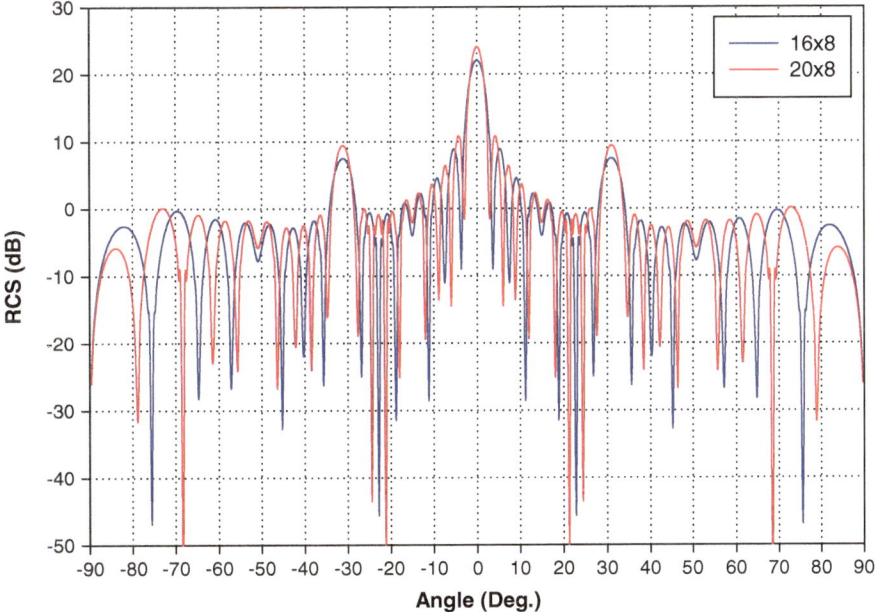

Fig. 10 Effect of array size on broadside RCS pattern of rectangular dipole array with parallel-feed network

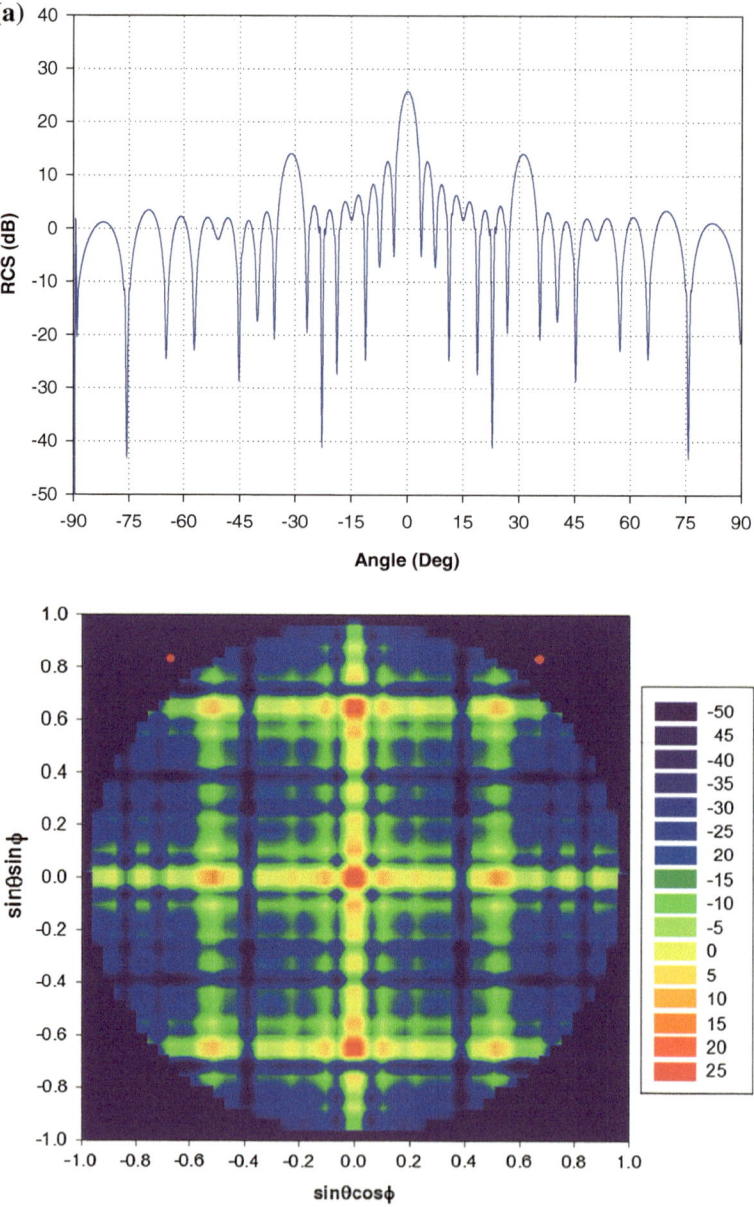

Fig. 11 Effect of beam scanning on RCS pattern of 16×10 planar parallel-fed dipole array. **a** $\theta_s = 0°$. **b** $\theta_s = 45°$. **c** $\theta_s = 60°$

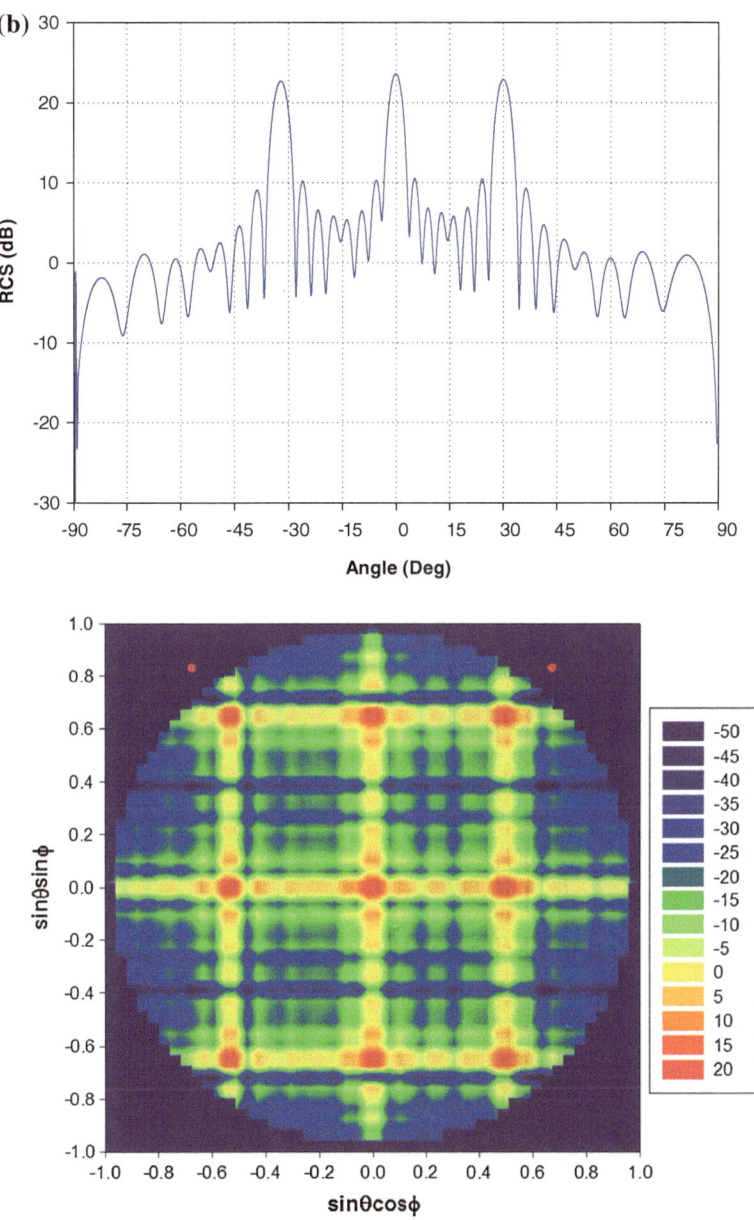

Fig. 11 (continued)

Figure 12 shows the broadside RCS pattern of a 16 × 16 dipole array with different terminating load impedances (50, 90, 150, and 180 Ω). The characteristic impedance of 50 Ω is considered. It is apparent that the array RCS is highest when

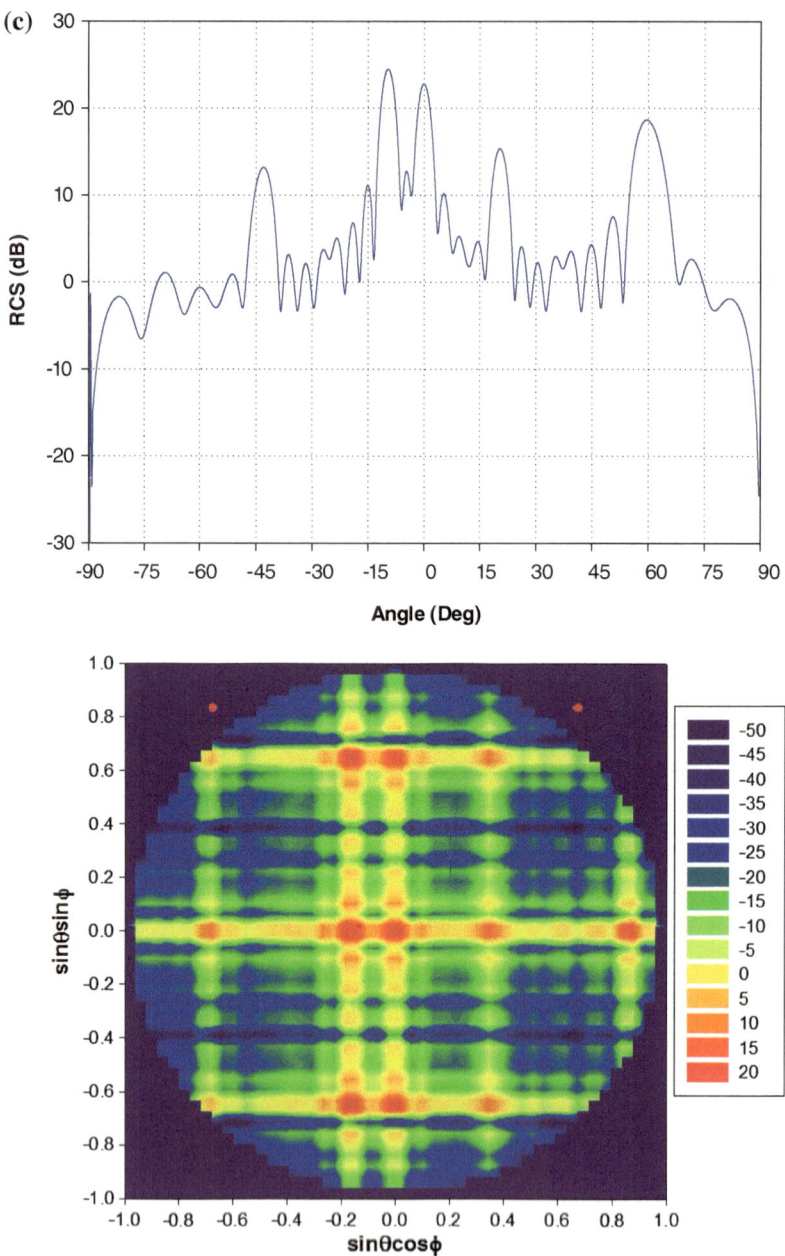

Fig. 11 (continued)

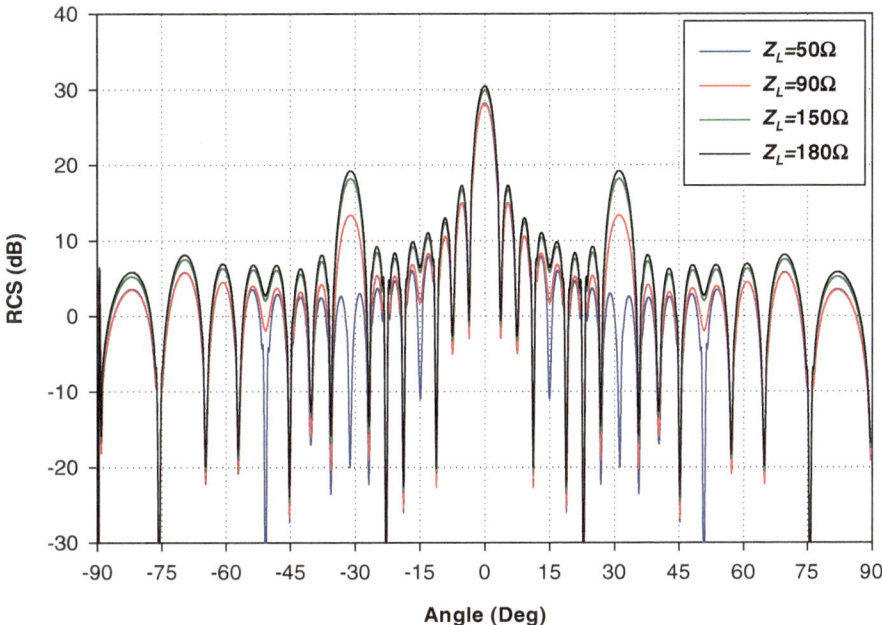

Fig. 12 Effect of terminating load impedance on RCS pattern of 16 × 16 planar dipole array

the coupler is terminated by 0 Ω (i.e. short circuited). If the terminating load impedance is increased to 90 Ω, the RCS value decreases at specular lobe, lobes due to coupler mismatches, and Bragg's lobes. However this trend has limiting value. On further increase in load impedance i.e. to 150 Ω and then to 180 Ω, the RCS value at lobes increases.

Another case of a 16 × 10 side-by-side dipole array is considered for analyzing the role of terminating load impedance on array RCS. Figure 13 shows the broadside RCS pattern of a 16 × 10 side-by-side dipole array with different terminating load impedances. It is visible that the array RCS is highest when the coupler is terminated by 50 Ω. If the value of load impedance is increased to 100 Ω, the RCS value decreases at specular lobe, lobes due to coupler mismatches, and Bragg's lobes. However this trend in RCS has limiting value. On further increase in load impedance i.e. to 125 and 180 Ω, the RCS increases. This makes the terminating load as an important design parameter towards the RCS optimization.

As a next parameter, the effect of coupler level on broadside RCS pattern is studied for a 32 × 32 dipole array (Fig. 14). Figure 14 presents both rectangular and contour plots. The characteristic impedance and the load impedance are taken as 50 and 150 Ω respectively. It can be observed from rectangular as well as contour plots that when scattering till second level couplers is considered for RCS estimation of 32 × 32 parallel-fed dipole array, additional lobes representing scattering in second level of couplers are visible in RCS pattern. These extra lobes are absent when scattering till first level couplers is considered.

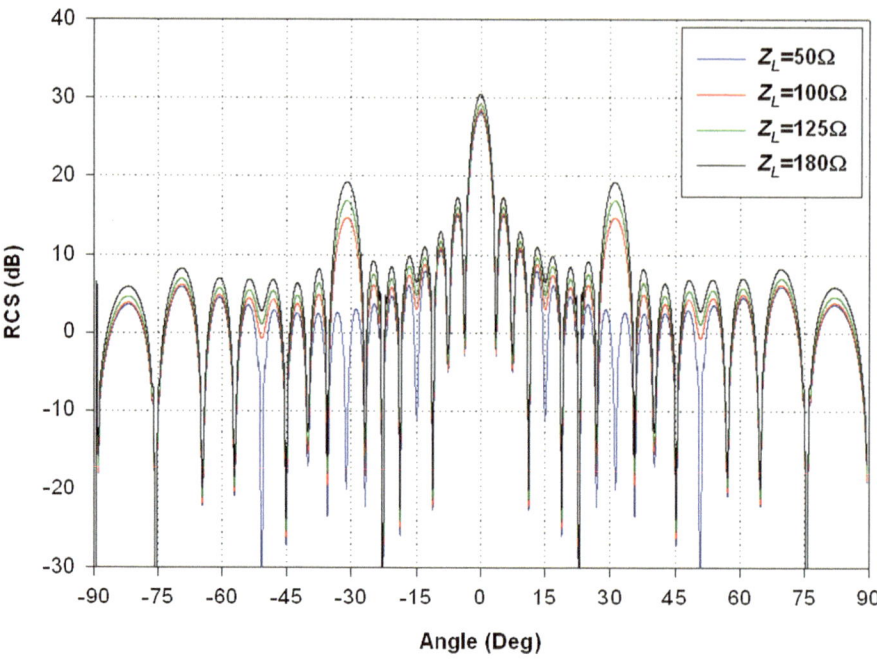

Fig. 13 Effect of terminating load impedance on RCS pattern of 16 × 10 planar side-by-side dipole array

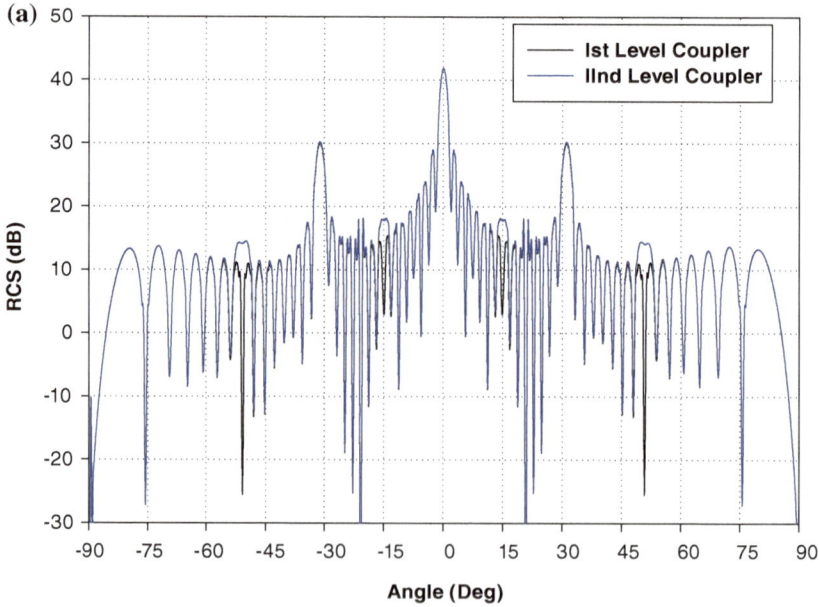

Fig. 14 Broadside RCS pattern of 32 × 32 parallel-fed dipole array till first and second level couplers. **a** Rectangular plot. **b** Contour plot

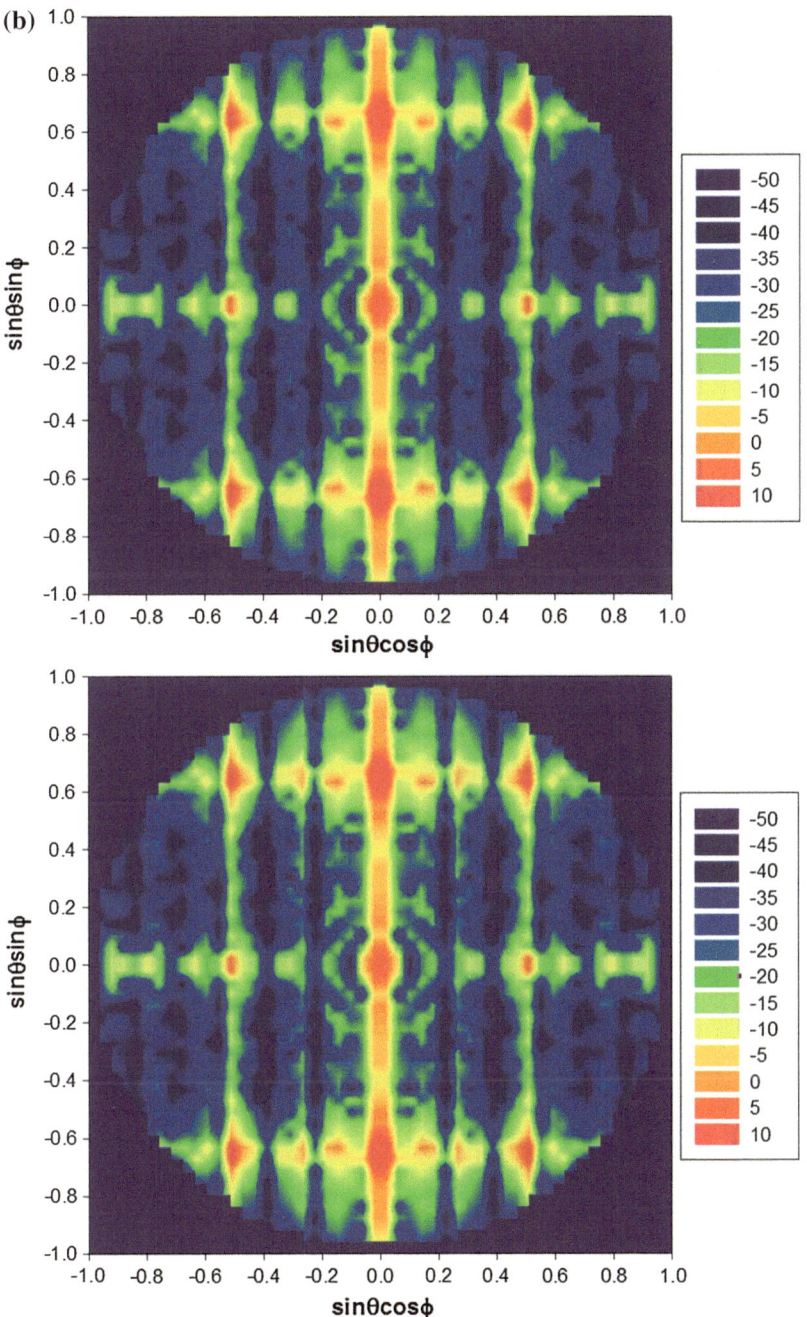

Fig. 14 (continued)

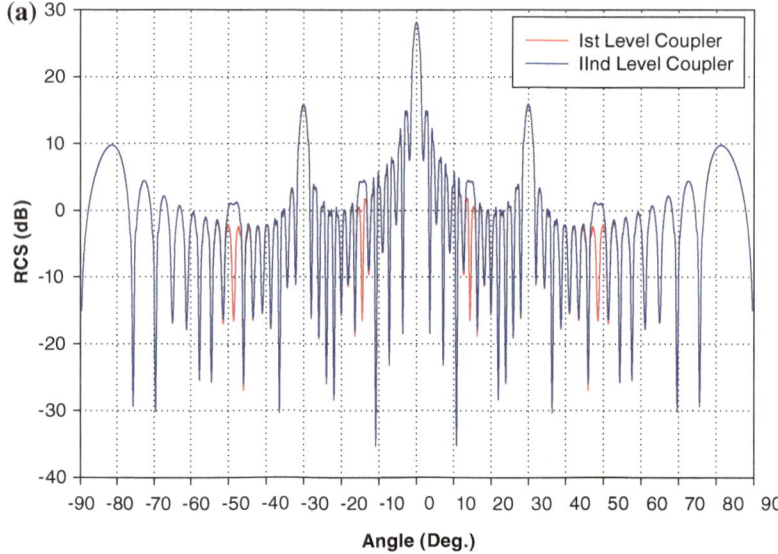

Fig. 15 Effect of coupler level on broadside RCS pattern of 32 × 10 parallel-fed dipole array till first and second level couplers. **a** Rectangular plot. **b** Contour plot

In order to further analyze the effect of coupler levels on overall array RCS, a rectangular 32 × 10 dipole array is considered. The inter-element spacing is taken as 0.5λ along x- and y-directions. The characteristic impedance and load termination are 50 and 90 Ω respectively. The length and radius of the dipole element are taken as 0.5λ and 0.01λ respectively. The extra small lobes due to second level couplers may be noticed even for rectangular dipole array (Fig. 15). The trend in RCS variation is same as in square dipole array. However the difference in contours and their levels is more prominent in rectangular dipole array than the square array (Fig. 14) when scattering is considered till first level of couplers and till second level of couplers in parallel feed network.

Next the role of inter-element spacing in RCS pattern of planar dipole array is analyzed. Figure 16 presents the broadside RCS pattern of 16 × 10 dipole array for different inter-element spacing along x-axis. The mutual coupling between antenna elements are taken care of. The antenna elements are excited using uniform amplitude distribution.

If the inter-element distance along the x-axis is increased from 0.35λ to 0.75λ, the mainlobe width and the specular lobe level decreases. The position of specular lobe and lobes due to coupler mismatches remains same. However the RCS level of the lobes in case of $d_x = 0.484\lambda$ is minimum making it as preferred choice of inter-element spacing. The corresponding contour plots are shown in Fig. 17, which further demonstrates the role of spacing between the dipole antenna elements on array RCS. The RCS value for $d_x = 0.484\lambda$ proves to be the optimum inter-element spacing towards low array RCS whether it is linear or planar dipole array. Likewise

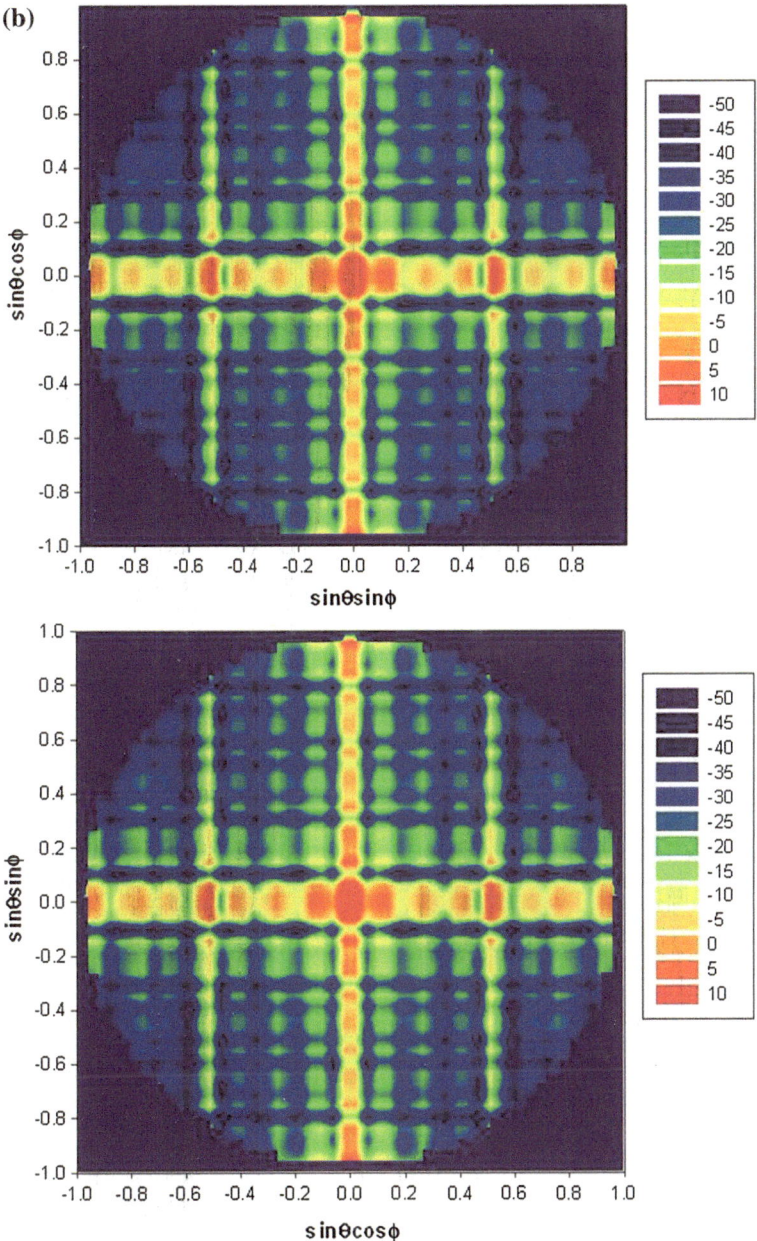

Fig. 15 (continued)

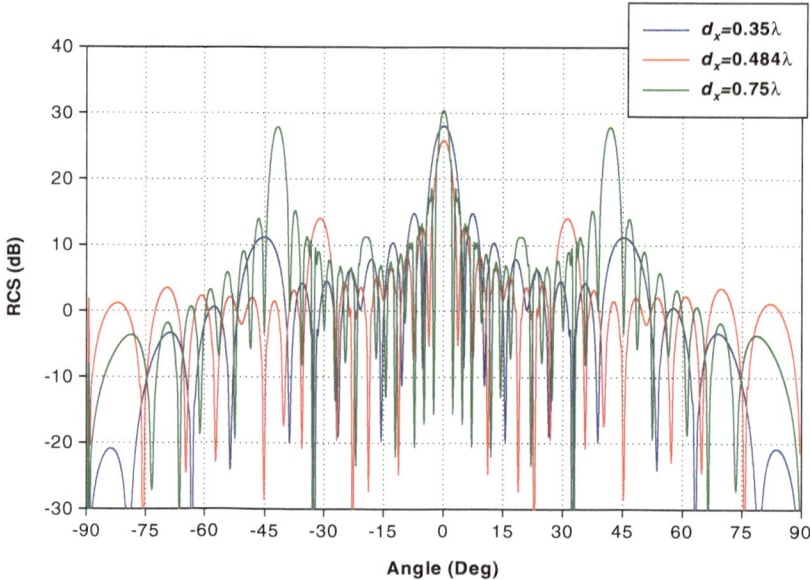

Fig. 16 Effect of inter-element spacing (d_x) on RCS pattern of 16 × 10 planar dipole array

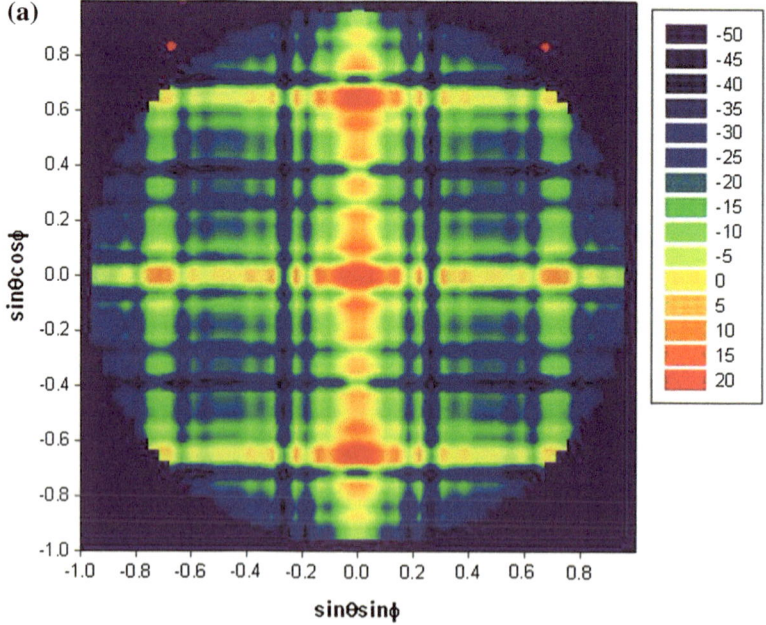

Fig. 17 Contour plot of broadside RCS pattern of 16 × 10 parallel-fed dipole array. **a** $d_x = 0.35\lambda$, **b** $d_x = 0.484\lambda$, **c** $d_x = 0.75\lambda$

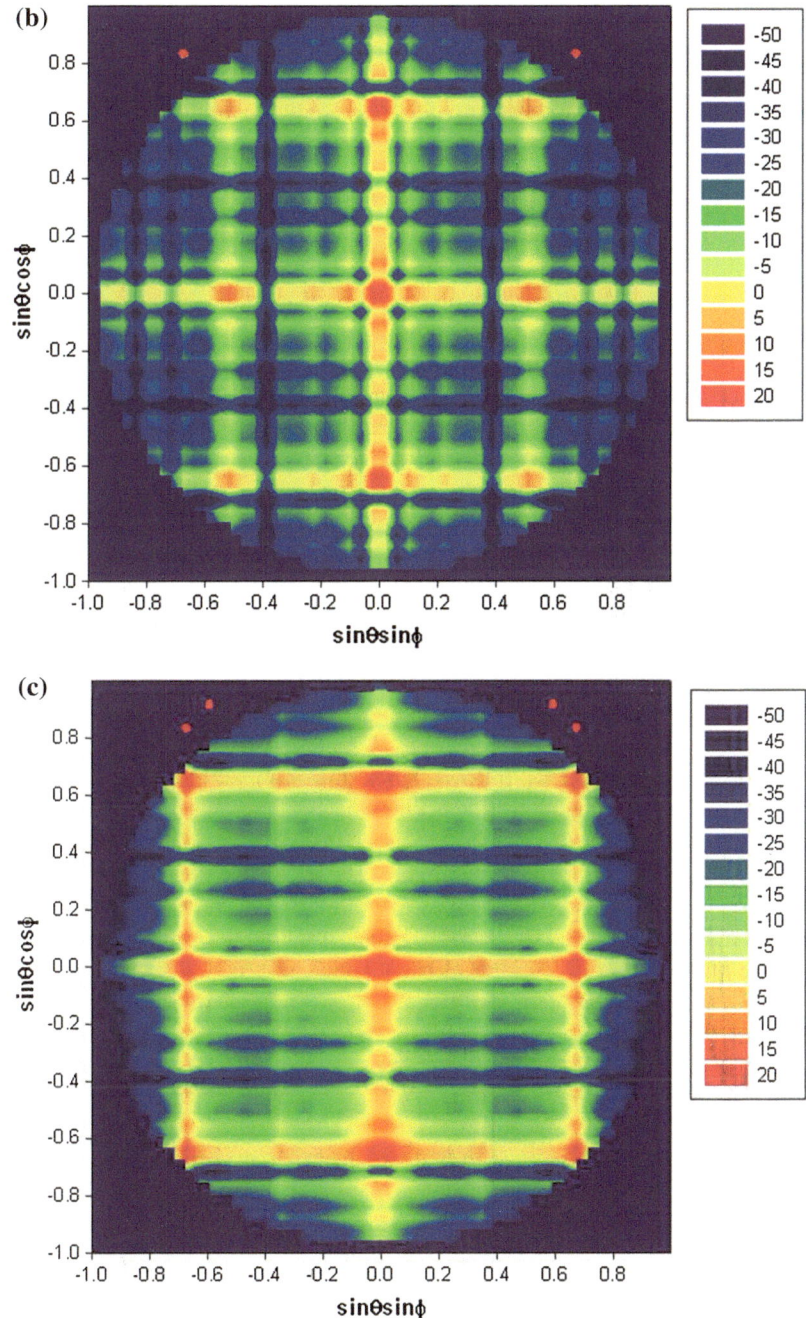

Fig. 17 (continued)

choosing other parameters such as load impedance, phase shifters, array configuration, design parameters of antenna element can facilitate the RCS control of phased arrays. However it may be constrained by other factors such as platform over which the array is mounted, neighboring systems, polarization etc.

6 Conclusion

This book presents the estimation of the RCS of parallel-fed uniform linear and planar dipole array in the presence of mutual coupling effect. An approximate method is used to derive the RCS of dipole array in terms of array factor. The phase terms are neglected. The RCS pattern is compared for the cases when scattering is considered till first level and till second level couplers in a parallel feed network. The signal enters the array system and travels through radiating elements, phase shifters, couplers before reaching the receive port. The scattered field at each level of the feed network is expressed in terms of the reflection and the transmission coefficients, owing to the impedance mismatches at various junctions of feed network. These individual scattered fields are coherently summed to obtain the total RCS of planar dipole array. The mutual coupling is included in terms of mutual impedance while calculating scattered field. The effect of varying the design parameters like number of elements, inter-element spacing, array configuration, beam scan angle and terminating load on the RCS pattern is studied. The inferences made in this parametric study of array RCS of parallel-fed dipole phased antenna array can be summarized as follows:When the number of antenna elements is increased, the level of major lobes (in linear array) and contours (in planar array) also increases. Moreover increase in the number of array elements makes the lobes or contours in the RCS pattern comparatively narrower, sharper and prominent. This variation in the level of the lobes or contours may be explained in terms of physical (and hence the effective) area of dipole array which depends on the number of array elements for a given inter-element spacing.

It is noted that the variation in the number of dipole elements does not alter the position of major lobes or contours in the array RCS. Specifically, in linear configuration, the number of sidelobes increases with increase in array size. In planar dipole array, the variations in contour levels do not provide clear inference towards the number of sidelobes.

When inter-element spacing is varied, no change in the location of specular lobe and other prominent lobes for both linear or the planar dipole arrays is observed. This may be due to increase in the effective aperture area of the array. However this is not the case in the level of lobes or contours. With increase in the spacing between dipole antennas, the number of lobes (in linear array) or contours (in planar array) also increases.

The beam scanning alters the RCS pattern of dipole array significantly. The level and position of lobes changes according to the scan angle. This is due to the fact that the lobes and the contours associated with the scattering within the array system is controlled by the phase shifter settings. However, the lobes or the contours due to the scattering in array system beyond the phase shifters scan along the beam. The RCS contributions due to the radiating elements (dipoles) and the components before phase shifters remain fixed. This is evident from the position of specular lobe in linear array and the main contour at (0, 0) in planar dipole array.

Further, the role of couplers is also significant in overall array RCS. Extra lobes arise due to scattering in higher level of couplers. The spatial separation between the lobes for different levels of couplers depends on their relative separation. Therefore it may be inferred that more the number of coupler levels in the feed network, more number of lobes or contours appear in the RCS pattern.

It may be noted that the effects observed on the array RCS due to the variation of design parameters are independent and uncorrelated. For instance, on varying the number of antenna elements and hence array size, only the level of the major lobes and the number of minor lobes in the array RCS pattern change. In contrast, the number of coupler levels decides the number of lobes due to the impedance mismatches in couplers. The level of the lobes remains unchanged. However on varying the design parameters in combination there may be mixed effect on the array RCS pattern.

The approximate model employed proves to be an efficient method for RCS estimation of dipole phased array. It has less computational complexity as compared to the conventional method of tracing signal as it travels through the antenna system. This book provides an insight of role of various design parameters for RCS control of phased arrays. These parameters if appropriately chosen contributes significantly towards optimization of array RCS.

References

Balanis, C.A. 2005. *Antenna theory, analysis and design*, 1117. Hoboken, New Jersey: Wiley. ISBN: 0-471-66782-X.

Elliot, R.S. 2005. *Antenna theory and design*, 594. Singapore: IEEE Press, Wiley (Asia). ISBN: 981-253-1947.

Jenn, D.C. 1995. *Radar and laser cross section engineering*, 476. Washington: AIAA Education Series. ISBN: 1-56347-105-1.

Jenn, D.C., and V. Flokas. 1996. In-band scattering from arrays with parallel feed networks. *IEEE Transactions on Antennas and Propagation* 44: 172–178.

Sneha, H.L., H. Singh, and R.M. Jha. 2012. *Radar cross section (RCS) of a series-fed dipole array including mutual coupling effect*, 36. Bangalore, India: CSIR-National Aerospace Laboratories. Project Document PD AL 1222.

Sneha, H.L., H. Singh, and R.M. Jha. 2013a. Scattering analysis of unequal length dipole array in the presence of mutual coupling. *IEEE Antennas and Propagation Magazine* 55(4): 333–351.

Sneha, H.L., H. Singh, and R.M. Jha. 2013c. *Back-scattering cross section of a parallel-fed dipole array including mutual coupling effect*, 51. Bangalore, India: CSIR-National Aerospace Laboratories. Project Document PD CEM 1306.

Sneha, H.L., H. Singh, and R.M. Jha. 2013b. Scattering analysis of a compact dipole array with series and parallel feed network including mutual coupling effect. *International Journal of Antennas and Propagation* 516946: 12.

Sneha, H.L., H. Singh, and R.M. Jha. 2014. Analytical estimation of radar cross section of arbitrary compact dipole array. *The Applied Computational Electromagnetics Society Journal* 29(9): 11.

About the Book

In this book, the RCS of a parallel-fed linear and planar dipole array is derived using an approximate method. The signal propagation within the phased array system determines the radar cross section (RCS) of phased array. The reflection and transmission coefficients for a signal at different levels of the phased-in scattering array system depend on the impedance mismatch and the design parameters. Moreover, the mutual coupling effect in between the antenna elements is an important factor. A phased array system comprises of radiating elements followed by phase shifters, couplers, and terminating load impedance. These components lead to respective impedances towards the incoming signal that travels through them before reaching receive port of the array system. In this book, the RCS is approximated in terms of array factor, neglecting the phase terms. The mutual coupling effect is taken into account. The dependence of the RCS pattern on the design parameters is analyzed. The approximate model is established as an efficient method for RCS estimation of phased arrays. This book presents a detailed formulation of approximate method to determine RCS of phased arrays, which is explained using schematics and illustrations. This book should help the reader understand the impinging signal path and its reflections/transmissions within the phased array system.

© The Author(s) 2016   43
H. Singh et al., *RCS Estimation of Linear and Planar Dipole Phased Arrays:*
Approximate Model, SpringerBriefs in Computational Electromagnetics,
DOI 10.1007/978-981-287-754-3

Author Index

© The Author(s) 2016
H. Singh et al., *RCS Estimation of Linear and Planar Dipole Phased Arrays:
Approximate Model*, SpringerBriefs in Computational Electromagnetics,
DOI 10.1007/978-981-287-754-3

Subject Index

© The Author(s) 2016   47
H. Singh et al., *RCS Estimation of Linear and Planar Dipole Phased Arrays:*
Approximate Model, SpringerBriefs in Computational Electromagnetics,
DOI 10.1007/978-981-287-754-3